CATASTROPHIC THINKING

science · culture

A series edited by Adrian Johns

CATASTROPHIC THINKING

*Extinction and the Value of Diversity
from Darwin to the Anthropocene*

David Sepkoski

THE UNIVERSITY OF CHICAGO PRESS

CHICAGO AND LONDON

The University of Chicago Press, Chicago 60637
The University of Chicago Press, Ltd., London
© 2020 by The University of Chicago
All rights reserved. No part of this book may be used or reproduced in
any manner whatsoever without written permission, except in the case
of brief quotations in critical articles and reviews. For more information,
contact the University of Chicago Press, 1427 E. 60th St., Chicago, IL 60637.
Published 2020
Printed in the United States of America

29 28 27 26 25 24 23 22 21   2 3 4 5

ISBN-13: 978-0-226-34861-2 (cloth)
ISBN-13: 978-0-226-35461-3 (e-book)
DOI: https://doi.org/10.7208/chicago/9780226354613.001.0001

Published with support of the Susan E. Abrams Fund

Library of Congress Cataloging-in-Publication Data

Names: Sepkoski, David, 1972– author.
Title: Catastrophic thinking : extinction and the value of diversity / David Sepkoski.
Other titles: Science.culture.
Description: Chicago : University of Chicago Press, 2020. | Series:
    Science culture | Includes bibliographical references and index.
Identifiers: LCCN 2020004118 | ISBN 9780226348612 (cloth) |
    ISBN 9780226354613 (ebook)
Subjects: LCSH: Extinction (Biology) | Biodiversity.
Classification: LCC QH78 .S475 2020 | DDC 576.8—dc23
LC record available at https://lccn.loc.gov/2020004118

♾ This paper meets the requirements of ANSI/NISO Z39.48–1992
(Permanence of Paper).

In Memory of David M. Raup (1933–2015)

For Sid and Ella, with apologies for the world we've left you

CONTENTS

INTRODUCTION:
WHY EXTINCTION MATTERS

If you were a dinosaur unlucky enough to be living in the region that is now the Yucatán Peninsula one fateful day 65 million years ago, you would have been startled by a blinding flash of light across the entire sky. An instant later, life on earth was changed forever. The flash of light was caused by a meteorite (or a comet) the size of Mount Everest — some ten to fourteen kilometers in diameter (10^{14} tons) — entering the earth's atmosphere travelling between thirty and seventy kilometers per second. The flash was caused as the intense speed and pressure of the asteroid heated the atmosphere underneath it to temperatures four to five times hotter than the sun. The impact itself was almost unimaginably devastating: the energy released was equivalent to one hundred million megatons of TNT, or roughly ten thousand times the combined destructiveness of the entire nuclear arsenal at the height of the Cold War. The first effect was to bury a crater almost forty kilometers deep into the earth's crust, simultaneously ejecting some one hundred cubic kilometers of earth into the atmosphere in a twenty-thousand-degree fireball that reached into space. All living things within several hundred kilometers were simply vaporized. Earthquakes of magnitude 12 or 13 rippled outward from the impact, violently buckling the earth's crust hundreds of meters into the air. This seismic activity triggered a massive tsunami, perhaps a kilometer high, that swept across the Gulf of Mexico and struck the coastline with enough force to travel twenty kilometers inland.

But this was only the beginning. As the material ejected from the impact reentered the atmosphere, it fell in a fiery rain across the globe, triggering wildfires that engulfed entire continents. The combined soot and dust in the atmosphere blocked out the sun for several months, enveloping the earth in near-total darkness. Photosynthesis stopped completely. When the rains finally came to wash away the soot, they carried deadly nitric acid formed when the superheated atmosphere bonded nitrogen, oxygen, and hydrogen molecules. Even after the skies cleared, the enormous amounts of CO_2 released when the asteroid impacted the limestone layer in the earth's crust remained in the atmosphere, causing a massive greenhouse effect that lasted for thousands of years. The aftermath of this event saw the total extinction of some 70 percent of all living species on earth—including, of course, the dinosaurs.

This scenario is part of a hypothesis advanced by a team of scientists led by the father-and-son duo of Luis and Walter Alvarez in 1980. It was based on their investigations of an anomalous layer of iridium, an element not commonly found on earth, at the boundary of the Cretaceous and Tertiary periods, roughly 65 million years ago. It set off a flurry of scientific activity that garnered international media attention for a decade, and fundamentally changed the way we understand the nature of extinction. Other spectacular claims followed: a team of paleontologists announced that the K-T extinction (K, the symbol for Cretaceous, refers to the characteristic chalk—*Kreide* in German—found in many deposits) was only one of at least five major extinction events during the past 250 million years, and not even the biggest one at that. To this they added the startling conclusion that these extinction events appeared to be spaced regularly in time, occurring every 26 million years.[1] A group of geophysicists and astronomers contributed to this finding by hypothesizing the existence of a mysterious companion star, which they dubbed "Nemesis," that traveled in an eccentric orbit around the solar system, periodically disturbing comets and raining death on the earth.

These sensational claims garnered enormous public attention, no doubt in part because they proposed a solution to the long-standing mystery of the demise of the dinosaurs. But the extinction hypotheses also tapped into a broad public awareness—and paranoia—about impending nuclear apocalypse that was fed by movies like *On the Beach,*

The Day After, Threads, and *Testament,* which realistically depicted not only the horror of nuclear Armageddon but also the chilling prospect of the "nuclear winter" that would follow. It is no accident, in other words, that the public was fascinated with scientific hypotheses about dooms- day scenarios from the past at a time when many believed humanity was on the edge of its own nuclear self-immolation.

At the same time, during the mid-1980s, another major scientific movement was gathering momentum and would come, in the next de- cades, to grip our attention even more strongly. This was the burgeoning awareness that the earth faced an impending "biodiversity crisis." Since the 1970s, a number of scientists had been giving voice to an increas- ing sense of alarm about the rapid depletion of worldwide ecosystems and the potentially permanent loss of many species and habitats. While a long history of conservation efforts in ecology and biology certainly contributed to this awareness, there was something genuinely new in the way this public discussion focused not just on protecting one or a few individual species or habitats, but rather on preserving the entire diverse global ecosystem itself. In this sense, biological diversity was identified as an inherent property of healthy ecosystems, and as a value in itself.

Another novel feature of the emerging biodiversity movement was the specter of catastrophic mass extinction. This was precisely what paleontologists and geologists had become interested in as a driving force in historical ecological change, and it came to haunt news reports, documentaries, scientific articles, and popular books championing bio- diversity. Mass extinction was an idea that had long been associated with the "catastrophism" of nineteenth-century scientists like Georges Cuvier, who had argued that the earth's history has been shaped by peri- odic drastic "revolutions" that have altered both the physical and the biological makeup of the globe. But for sober Victorian naturalists like Charles Darwin and the geologist Charles Lyell, to whom this sounded a bit too much like biblical geology, a picture of geological history in which changes took place very slowly and gradually made much more sense. So, for the next century or so, geologists and paleontologists were careful to avoid the subject of catastrophic mass extinctions—meaning that the Alvarez hypothesis and other paleontological investigations of extinction represented a fairly radical "new catastrophism" movement that had been gaining popularity since the late 1960s. The language of the new catastrophism was evident in the biodiversity rhetoric right from the start. For example, the ecologist Norman Myers's influential 1979 book *The Sinking Ark* characterized current rates of species extinction as being potentially more disastrous than the event that killed the dinosaurs, and warned that it was "happening within the twinkling of an evolutionary eye" (Myers 1979, ix).

Consciousness-raising among scientists about threats to biological diversity reached a critical mass in 1986, when the entomologist and ecologist E. O. Wilson teamed up with the botanist Walter G. Rosen to host a "National Forum on BioDiversity" in Washington, DC. This event, cosponsored by the National Academy of Sciences and the Smithsonian Institution, was the first major interdisciplinary conference on the biological diversity crisis, and brought together major figures in biology, ecology, paleontology, economics, and public policy. It garnered a significant amount of media attention—in both the scientific and the popular press—and is widely credited with launching biodiversity preservation as an organized movement.

From the very beginning, extinction was central to the way the organizers perceived the "crisis." As Wilson put it in the introduction to the companion volume to the conference, "The current reduction of diversity seems destined to approach that of the great natural catastrophes at the end of the Paleozoic and Mesozoic eras—in other words, the most extreme in the past 65 million years."[2] Wilson's contribution—like many of his later writings on the subject—made frequent references to paleontological studies of mass extinctions, which he used to establish parameters to distinguish between "normal" and "extraordinary" levels of extinction. This strategy has been picked up in nearly all subsequent discussions of biodiversity, so much so that the current crisis is often referred to as the "Sixth Mass Extinction," in reference to the five major mass extinctions identified by paleontologists in the geological past. This Sixth Extinction concept has achieved wide cultural currency, in part because of the success of the Pulitzer Prize–winning 2014 book by Elizabeth Kolbert *The Sixth Extinction: An Unnatural History*. It has also influenced a broad array of current discussions about the impact of anthropogenic climate change and associated environmental crises for the future of human society associated with the so-called "Anthropocene" concept. Debates about the future of humanity itself, then, are closely tied to understandings of mass extinctions and environmental catastrophes in the deep history of the earth.

The problem this book addresses is how the development of scientific and cultural understandings of extinction over the past two hundred years have influenced—and been shaped by—the way Western culture has understood the health and stability of its current society and its prospects for the future. One of the central components in these discussions has been the way Westerners have (or have not) valued and appreciated diversity. Diversity is now widely regarded as an essential biological and cultural resource, and it has become closely tied to the sense of fragility imperiling both the natural and social worlds—so much so that, during the 1990s and the 2000s, the United Nations produced resolutions calling for the protection of both biological and cultural diversity as essential human "resources." This investment of diversity with the language of resource and endangerment, however, only emerged as part of a long historical development whose history this book will narrate. Ultimately, it was because of a set of specific, contingent, and fairly recent historical circumstances that we learned to "think catastrophically" about the threats facing both our natural world and our human future.

The central argument of this book is that the way we understand the relationship between humans and the rest of the natural world—and how we conceive of ecological relationships, geological processes, and evolutionary dynamics—shapes the kind of futures we can imagine for our species. It informs the kinds of scientific questions we ask, the political and technological ambitions we pursue, our anxieties about the present and the future, and the basic values that guide our interactions with one another and with the organisms with whom we share the planet. The word "imagination" is an important concept in this book. As the legal scholar and cultural observer Jedidiah Purdy has recently put it in his excellent book *After Nature*, "What we become conscious of, how we see it, and what we believe it means—and everything we leave out—are keys to imagining the world. . . . Imagination also enables us to do things together politically: a new way of seeing the world can be a way of valuing it—a map of things worth saving, or of a future worth creating" (Purdy 2018, 7).[3] The complex web of values and beliefs associated with extinction at any given historical period forms what I will call, to use an academic term of art, an extinction "imaginary."[4] The way we understand extinction—the extinction imaginary of any given time—is ultimately tied to the way we conceive of the basic stability and security of the continued existence of our own species.

Extinction imaginaries are co-constructed both by contemporary scientific theories about extinction and by broader cultural attitudes and values about social progress, technological innovation, ethical responsibilities towards nature and our fellow humans, and the nature of history itself.[5] Scientific understanding of extinction has changed quite dramatically over the past two hundred years, as have these other aspects of Western cultural belief, and it is my adamant position that these changes have been linked and are mutually reinforcing. Historians have long since given up, for the most part, debating whether science is a product of human culture; that scientists and the science they produce are conditioned by, and in turn contribute to, wider social, political, and cultural values and beliefs will be treated as a basic assumption of this book. For any reader with doubts on this score, this book will also amply document that this is the case. But my larger argument is that the extinction imaginary, as a *particular* example of the co-construction of scientific and cultural values, has shaped how we understand ourselves, our history, and our future in very specific and important ways.

I first became interested in the history of extinction nearly twenty years ago, and I eventually wrote a long book documenting the history of paleontological approaches to studying the patterns of life's history, including the study of diversification and extinction over hundreds of millions of years.[6] But my broader interest in extinction goes back much further. My father, Jack Sepkoski (fig. 0.1), was a paleontologist who was centrally involved in a "renaissance" of extinction research during the 1970s and 1980s (he died in 1999 at the age of fifty, while I was still in graduate school). Growing up with him, I was fascinated by the strange creatures and landscapes of the distant past that he would describe to me, and was haunted by the notion that the magnificent and fearsome

FIGURE 0.1 The author's father, J. John "Jack" Sepkoski Jr., examining the Cretaceous-Tertiary boundary in an outcropping outside Los Alamos, New Mexico, in 1986. Photograph by Karl Orth. Personal collection of the author.

dinosaurs could have been wiped out in one terrible, catastrophic instant. As a child of the late 1970s and 1980s, I found that this resonated
deeply with my own anxieties about the fate of our own species, and I still
vividly remember sitting with my parents watching the ABC television
movie *The Day After*, which dramatically and realistically depicted the
aftermath of a nuclear war. I would frequently experience nightmares
in which I was awakened by a flash of light and looked out my bedroom
window to see a mushroom cloud silently rising from downtown Chicago—which somehow, in my mind, connected to the new story about
the fate of the dinosaurs about which my father was suddenly being
interviewed for magazine articles and science documentaries. It didn't
occur to me at the time to wonder whether there was any connection
between the way scientists like my father understood mass extinction,
and the pervasive anxiety we all felt about nuclear war. But many years
later, having written widely about the scientific basis for these theories,
I came to be convinced that it was no accident that catastrophic mass
extinction became an object of scientific study and popular fascination
at precisely the moment when we imagined a similar fate for ourselves.
This is for me, then, a very personal history—but it is also personal for
all of us, in that it deals intimately not just with how we understand the
global past, but also with our very personal hopes and anxieties about
the future.

    This relationship is exemplified in, of all things, a 1984 newspaper
column whose author, Ellen Goodman, asked "whether every era gets
the dinosaur story it deserves." She explained that the dinosaurs of her
1950s childhood "were big, but their brains were small. The dinosaurs
couldn't adapt. Slowly they died out while humans, the adaptable,
thinking species, prospered." Now, however, we have learned that the
dinosaurs were merely "the victim of a climatic disaster, a cosmic accident," and that mass extinction has been a regular feature of the history
of life. What, Goodman mused, does this tell us about our science and
ourselves? She continued, "The scientists of the 19th century—a time
full of belief in progress—saw evolution as part of the planet's plan of
self-improvement. The rugged individualists of that century blamed the
victims for their own failure. Those who lived in a competitive economy
valued the 'natural competition of species.' The best man won." But

"surely we are now more sensitive to cosmic catastrophe, to accident. Surely we are more conscious of the shared fate of the whole species." Goodman concluded,

> Today the astronauts travel into space and report back that they see no national borders. Environmentalists remind us that the acid from one nation's chimneys rains down on another. Most significantly, another group of scientists warns us that a nuclear war between two great powers would bring a universal and wintry death. . . . In that sense, the latest dinosaur theory fits us uncomfortably well. "Our" dinosaurs died together in some meteoric winter, the victims of a global catastrophe. As humans, we fear a similar shared fate. The difference is that their world was hit by a giant asteroid while we—the large-brained, adaptable creatures who inherited the earth—may produce our own extinction" (Goodman 1984).

I think Goodman is exactly right: the stories a society tells itself about the fates of extinct prehistoric creatures have as much to do with that society's beliefs and values about the natural and social worlds of the present as they do with the past. During the nineteenth century, at a time when naturalists understood nature to be an essentially endlessly renewable resource, extinction was understood to be nature's way of strengthening and improving itself by weeding out the unfit, and competition was celebrated as the source of natural progress. For the Victorians and their immediate descendants, dinosaurs were emblematic of the fate of all those who are unable to keep pace with a changing world, and who must therefore stand aside for those who could. The view of extinction held by Darwin and other nineteenth-century naturalists was that extinction is (1) slow and gradual, (2) reciprocally balanced by the replenishment of new species, and (3) in some sense progressive. That is, by reflecting the "fair" outcome of natural competition, it contributes to the robustness of living ecosystems by weeding out "unfit" individuals or species.

Viewed from this perspective, diversity is an inherent and self-renewing property of the "economy of nature," and thus requires no special protection or independent valuation. As I will demonstrate in this book, this particular concept of extinction was central to a cultural

and political ideology—especially in Britain and the United States—
that supported imperialism and downplayed the value of protecting
species and peoples from threat of extinction. The crux of the matter is
that, in Victorian society and beyond, extinction was considered both
an inevitable and a progressive process, whether applied to humans or
to "lower" organisms. This view came from biology, but it is insepa-
rable from a broader set of cultural and political attitudes about race
and social progress. It certainly did not promote the active protection of
threatened peoples or organisms, nor did it celebrate intrinsic biological or cultural difference the way our society does today. "Diversity" was
not an independent value at that time in biology or culture, because it
had not been identified as something necessary for biological or cultural
stability. If anything, extinction was seen as a positive good: by removing the unfit, it acted for the betterment of species or "races." There was
no sense that when species or cultures disappeared, some valuable resource was being lost; rather, through the law-abiding process of natural selection, Nature was constantly improving her stock.

We now live in a society where cultural and biological diversity are
considered to be precious resources, and where threats to those resources are perceived from all directions. We fundamentally value diversity, as an inherent normative good, in a way that previous Western societies did not. This is partially due to the emergence in the
mid-twentieth century of a new understanding of extinction in which
(1) extinction is seen as a potentially catastrophic and irreversible process, (2) extinction is characterized explicitly in terms of its effect on
diversity, and (3) survival is no longer conceptualized as a "fair game"
in which extinction penalizes only those individuals and species who
"deserve" it. The transformation from the Victorian attitude to the one
broadly held today was a complex, drawn-out process. These ideas developed first in a scientific context of ecology and paleontology, but
have ramified outward to perceptions of cultural and linguistic diversity, and have become central to cultural valuations of diversity itself.
There is obviously an important sense in which scientists have themselves been influenced by changing cultural norms (paleontologists
were, after all, just as frightened by the specter of nuclear war as were
the rest of us), and this book shows that the new understanding of ex-

tinction was made more acceptable by a cultural and political context in which nuclear proliferation and environmental catastrophe were looming specters.

Two central scientific features of this transformation were the development of a new ecological understanding of what "balance" meant in nature, which began to take shape in the 1920s and '30s, and the emergence of what has been called the "new catastrophism" in paleontology during the 1970s and '80s. In the late nineteenth century, biologists generally regarded extinction to be a problem that was "solved." One can see just how much this view had changed 100 years later in the comment of David Raup, one of the most prominent extinction theorists in paleontology, in a letter to a colleague: "I am becoming more and more convinced that the key gap in our thinking for the last 125 years is the nature of extinction" (Raup to Schopf, January 28, 1979).[7] What Raup meant was that paleontology—and biology more generally—had no adequate theory for the causes and consequences of extinction. Here Raup laid the blame directly at Darwin's doorstep: by focusing exclusively on natural selection and competitive replacement as the cause of extinction, Darwin's view effectively presented a tautology with little explanatory value, where "the only evidence we have for the inferiority of victims of extinction is the fact of their extinction" (Raup 1991, 17).

Whereas Darwin himself believed that levels of biological diversity remained constant over the history of life, what paleontologists who have since studied the fossil record found was a complex pattern of steep rises and sharp plummets in levels of diversity over the past 500 million years. Through work carried out by Raup and other paleontologists during the 1970s and 1980s, it became apparent that major catastrophic mass extinctions had played a key role in perturbing the history of life many times. These mass extinctions were episodes that typically lasted no more than a few million years, but where anywhere from 50 to 95 percent of all existing species died out. In 1984, Raup and my father (who were colleagues at the University of Chicago) argued that, remarkably, these mass extinctions appear to follow a regular periodicity, occurring roughly every 26 million years. The major evolutionary interpretation this suggested was that these events could not be explained as the product of natural selection alone; they were catastrophic episodes that effectively "reshuffled the deck" for evolution, wiping out long-standing groups (like the dinosaurs), and ushering new ones (such as the mammals) to evolutionary prominence. If the significance of these mass extinctions was to be credited, this presented an entirely new view of extinction: while normal or "background" extinctions probably occurred as slowly and constantly as Darwin had held, a significant mechanism in the history of life and diversification was events that appeared to follow no Darwinian rules of selectivity, in which entire taxonomic groups disappeared through no "fault" of their own. A central message of this new interpretation was that life on earth has been much more dynamic—and its continuation more tenuous—than anyone previously had imagined. The Raup-Sepkoski extinction work happened to coincide with the Alvarez team's discovery of evidence that the dinosaurs perished in a fiery cataclysm. The impact evidence was potentially the kind of nonselective trigger implied by the Raup-Sepkoski work, and it appeared to revise the earlier Darwinian logic of extinction in dramatic ways. Raup has most succinctly reduced the problem to a question of whether extinction is caused by "bad genes or bad luck"—or, as he has put it, whether "the evolution of life [is] a fair game, as the survival-of-the-fittest doctrine so strongly implies" (Raup 1991). One upshot of this extinction work was the creation of a cottage industry in paleontological studies of mass extinction, and the legitimation of a new catastrophism. Another was that extinction was essentially redefined in terms of diversity: mass extinctions are recognized in the fossil record, explicitly, as those periods when diversity drops significantly in a short amount of time.

These findings created a sensation in the scientific community and the popular media, and for a short time paleontologists and geologists like Alvarez, Raup, and Sepkoski became minor media stars. Major magazines and newspapers, from *Time* and *Newsweek* to the *New York Times* and the *Washington Post*, gave the new impact-extinction theories front-page billing—and I vividly remember being both excited and nonplussed to see my own father, along with his colleagues whom I had known from casual backyard cookouts or boring academic parties, suddenly appearing in the national media. In fictional accounts, from science fiction novels to major Hollywood films, comet or asteroid impacts joined the more familiar theme of nuclear Armageddon as popular disaster scenarios, and "post-apocalyptic" became a pop-cultural buzzword. It is not difficult to understand why scientific theories about extinction would have caused such a stir: the dinosaurs have always been the most charismatic and popular prehistoric creatures, and their demise had remained an enigma for more than a century.

Another factor was the era of Cold War anxieties of nuclear annihilation and environmental catastrophe. If the dinosaurs could go, the idea went, then so could we humans. In fact, the model of "nuclear winter" that frightened the public during the mid-1980s was actually developed from climate models produced to estimate the atmospheric effects of the massive asteroid that likely struck 65 million years ago, thus making the juxtaposition of the fates of humanity and the dinosaurs more than merely metaphorical. It was at the height of the scientific and public interest in mass extinctions that in 1986 the biodiversity movement formally began. There were certainly earlier contributing factors: a long history of conservation efforts focused on preserving individual endangered species, for example. But there was something genuinely new about how the major proponents of biodiversity, people like Wilson and Norman Myers, mobilized interest in protecting not one or a few individual species or habitats, but the entire diverse global ecosystem itself. Biodiversity, in other words, helped make diversity a normative value.

The reasons for this are many and complex, but I will point to a few. In the first instance, ecologists began during the mid-twentieth century to better appreciate the fragility and interconnectedness of ecosystems. One couldn't focus on just the big, "charismatic" vertebrates and expect success; the insects and even microbes mattered, too, if one wanted to maintain healthy habitats. Second, a transition took place to a less romantic and more utilitarian environmentalist ethos than the one that had existed in the late nineteenth and early twentieth centuries. Conservation arguments increasingly tended to promote the economic, biomedical, and even ethical reasons for preserving all life, rather than those related to aesthetics and recreation. The biodiversity movement would follow this trend. Third, and quite simply, the pace of human depletion of the natural environment got a lot faster. Rain forest destruction, environmental pollution, sprawl, and a host of other problems had been accelerating since the demographic expansion of Western societies in the 1950s, making their consequences more and more apparent. Fourth, arguments began more frequently to be focused, from the 1960s and onward, on the danger of unforeseen consequences. While the utilitarian value of most species was unknown, the rapid pace of discovery in the pharmaceutical and other industries suggested that previously unknown or humble organisms might have great worth. Likewise, as the laws of ecological relationships were better understood, it occurred to many that irreparable harm might be done to fragile ecosystems before it was even realized. Diversity itself, in other words, became conceptualized as a vital resource.

Finally, biologists interested in conservation efforts became aware of the new science of extinction and its consequences, which gave them both a sense of the scope of the current crisis, and tools and data with which to predict its consequences. As Wilson put it in *The Diversity of Life*, "The laws of biological diversity are written in the equations of speciation and extinction" (Wilson 1992). Paleontological studies of mass extinction gave biodiversity proponents a set of arguments about the potential consequences—both for ecological recovery and in evolutionary terms—of allowing a "sixth extinction" to proceed unchecked. And extinction studies have helped silence the appeals to nature's ability to endlessly renew itself that characterized an earlier era of thinking. The fact that mass extinctions can and do occur, and that they have dramatic short and long-term consequences for diversity, has contributed a much greater sense of impending danger than was present in earlier conservation rhetoric. Extinction is no longer just something that we discuss when we are talking about the distant past, or about other species; it may be taking place now, and it may ultimately impact human beings. Extinction has become personal.

During the 1980s and 1990s the biodiversity movement brought about a new way of seeing and valuing natural diversity that embodied not only scientists' interpretations of empirical evidence, but also their "political, emotional, aesthetic, ethical, and spiritual feelings" (Takacs 1996). In other words, biological diversity came to be seen by scientists, policymakers, and the general public not just as important for ecological survival or medical and economic development, but as something "good" in itself. This shift occurred as many Western societies began to identify other kinds of diversity — cultural or linguistic, for example — as an inherent normative good. One of the clearest examples of the overlap between valuations of biological and cultural diversity is in the rhetoric used by the United Nations and UNESCO over the years to promote these ideals. A few years after the initial biodiversity conference was held in Washington, representatives from 150 nations took part in an "Earth Summit" held in Rio de Janeiro. The result was the United Nations Convention on Biological Diversity, which explicitly called attention to "the intrinsic value of biological diversity" (United Nations 1992). A decade later, UNESCO produced the Universal Declaration on Cultural Diversity, which framed cultural diversity in the same language of "resource" in which biological diversity was being presented: "The Declaration aims both to preserve cultural diversity as a living, and thus renewable treasure that must not be perceived as being unchanging but as a process guaranteeing the survival of humanity" (UNESCO 2002). The declaration went on to make the analogy between both forms of diversity explicit, stating in its article 1 that, "as a source of exchange, innovation and creativity, cultural diversity is as necessary for humankind as biodiversity is for nature."

This sense that cultural and biological diversity are not merely similar but actually manifestations of the same phenomenon can be seen in the emergence of a new term, "biocultural diversity," at around the same time. This conflation of biological and cultural diversity is nowhere more evident than in a UNESCO booklet published in 2003 titled *Sharing a World of Difference: The Earth's Linguistic, Cultural, and Biological Diversity*. This document defines biocultural diversity as "interlinkages between linguistic, cultural, and biological diversity," and asserts that "the diversity of life on Earth is formed not only by the variety of plant and animal species and ecosystems found in nature (biodiversity), but also by the variety of cultures and languages in human society (cultural and linguistic diversity)" (Skutnabb-Kangas, Maffi, and Harmon 2003). This cultural diversity can be thought of "as the totality of the 'cultural and linguistic richness' present within the human species," a quantity analogous to species and genetic richness in biology, and the world's six to seven thousand languages are "the total 'pool of ideas'" represented in human culture, all of which are threatened by a "linguistic and cultural extinction crisis" (Skutnabb-Kangas, Maffi, and Harmon 2003). But the conclusion the booklet reaches goes beyond mere analogical relationship: "Biological diversity and linguistic diversity are not separate aspects of the diversity of life, but rather intimately related, and indeed, mutually supporting ones," and "the extinction crises that are affecting these manifestations of the diversity of life may be converging also" (Skutnabb-Kangas, Maffi, and Harmon 2003). The central message is that, like biological diversity, cultural diversity is a resource for ensuring a healthy cultural "ecosystem" that, if lost, will be lost forever.

The rhetoric of diversity is certainly still contested; just ask any politician involved in legislation surrounding development of natural resources, or glance at the literature about linguistic or cultural diversity in public schools. The political left has become heavily invested in a particular formulation of the normative value of biological and cultural diversity, as have many politically conservative observers in opposing it as an example of "political correctness." Religious beliefs have also played a prominent role in valuations of diversity over the past two centuries, providing arguments for responsible stewardship as well as justification for exploitation (as evidenced, for example, by current religious conviction that the climate is in the hands of higher powers, a view recently expressed by Senator James Inhofe, who declared it outrageous to assume that humans could change what God had ordained).

As a society we do value diversity in many ways quite differently than did nineteenth-century Europeans and Americans, but we also struggle with what diversity is and what it means. While I do not claim that this book will definitively explain how the complex politics surrounding diversity have evolved, I do suggest that this broader examination of the way biological and cultural values surrounding extinction have developed over the past 200 years will shed light on some of the reasons why issues of diversity remain so contested. To want to preserve something, we must first perceive that it is threatened, and the emergence of a new—and personal—view of extinction has been central in underlining what kinds of threats we as a culture face.

1

THE MEANING OF EXTINCTION: CATASTROPHE, EQUILIBRIUM, AND DIVERSITY

Extinction exerts a powerful cultural fascination today. The extinction of particular species or groups of animals is often vested with romantic, tragic, and moral shading. We sometimes see the demise of the dinosaurs, for example, as an object lesson for our own hubristic species, or the helpless dodo as a symbol of the fragile innocence of nature, or the American bison as a reminder of the destructive potential of human expansion. But no matter how much we may regret or mourn the loss of particular species, we now know that extinction is a normal feature of the history of life, and part of the regular course of nature. Despite the centrality that extinction now has in our perceptions of nature, the recognition that extinction is a ubiquitous, even commonplace phenomenon represents a profound shift in scientific and cultural awareness of the tenuousness of life and the balance of nature that has taken place over the past two hundred years. In the late eighteenth century, for example, many naturalists doubted whether a species could ever become extinct at all, and when considered, extinction was treated as a rare phenomenon that took place only under dramatic, exceptional circumstances. Even when, by the mid-nineteenth century, scientists began to accept extinction as a more general feature of the history of life, it was widely held that nature maintained a constant equilibrium, where the loss of any one species would always be equally balanced by the appearance of a new one somewhere else.

The notion that nature is a self-regulating machine is one of the oldest ideas in Western philosophy. The notion of plentitude—that "all that can be imagined must be"—goes back at least as far as Plato and Aristotle, and with it the conception that nature persists in a maximum state of diversity, with no new living forms ever being created or lost. The concept of balance as the opposition of forces or elements was also a common theme, from the pre-Socratic philosophers through Aristotle, and Epicurean and Stoic authors also considered nature's stable balance to be an inherent feature of the world. In the Christian era, authors like Augustine of Hippo combined this essentially neo-Platonic idea with notions of divine beneficence, arguing that just as God had created every living thing required in a perfect world, neither would he suffer any class of organism to be created or destroyed. Later Christian authors such as Thomas Aquinas modified this view somewhat to preserve the freedom of God to act and to allow for the possibility of change, arguing, for example, that it was conceivable that God, in his infinite wisdom, might choose to add a species of organism or angel to make the universe even "more perfect," despite the apparent contradiction this might imply. However, the basic principle was that nature is preserved in a state of perfection, and that when change does take place it is precisely balanced so as to maintain that state.[1]

This idea persisted, more or less unaltered, into the beginnings of what historians consider the modern era of biology. A variety of authors in the seventeenth century discussed how the benevolent hand of God ensures that, despite the constant change observed in the organic world—incessant generation and corruption—a well-ordered and stable natural economy will obtain. This theme reached its early modern apex in the tradition of "physico-theology," a religious and scientific philosophy popular especially in England and exemplified by works by authors including the physician Walter Charleton, the experimental naturalist Robert Boyle, and the pioneering botanist and taxonomist John Ray. The central assumption of physico-theology was that God's actions, being rational, can be observed and understood using the tools of natural philosophy, and furthermore that God plays an active role in maintaining his orderly creation. This was no remote watchmaker

God: for Ray and others, God had an intimate concern with ensuring that nature was in a constant state of perfection, and to that end had designed each organism to play a role in a balanced natural economy. Ray explored these ideas in theoretical treatises like *The Wisdom of God Manifested in the Works of Creation*; but he also devoted his life to the study and cataloging of organisms, particularly plants, and was one of the great pre-Linnaean systematizers of the natural world.

Throughout the eighteenth and early nineteenth centuries, European writings about the economy of nature maintained these explicitly Christian providential overtones. In 1749 the great taxonomist Carolus Linnaeus published an influential essay titled *The Oeconomy of Nature*, where he argued, "By the Oeconomy of Nature are understood the all-wise disposition of the Creator in relation to natural things, by which they are fitted to produce general ends, and reciprocal uses" (Linnaeus 1762, 39).[2] Linnaeus was a deeply committed Lutheran for whom the study of nature was explicitly an exploration and celebration of the magnificence of God's creation. In this treatise Linnaeus dealt with a potential conflict: On the one hand, as a devout Christian he firmly believed that the perfection of creation meant that every conceivable natural place was filled. On the other, as a naturalist he was well aware that violence and death were inescapable. His solution was to conceive of the inevitable struggle among organisms as essential to nature's divinely ordained economy: "In order therefore to perpetuate the established course of nature in a continued series, the divine wisdom has thought fit, that all living things should contribute and lend a helping hand to preserve every species; and lastly, that the death and destruction of any one thing should always be subservient to the restitution of another" (Linnaeus 1762, 40). Importantly, while the destruction or death of any particular *individual* would have no net effect on the balance of nature, Linnaeus denied that God would ever suffer the extinction of an entire *species*. However, as we will see, this conception of the economy of nature would remain influential even after extinction was recognized as a genuine natural phenomenon. Furthermore, even when, by the mid-nineteenth century, most naturalists had abandoned explicitly religious justifications, ideas that ultimately stem from this Christian providentialist worldview continued to exert a strong influence. God may have ultimately been excluded from the system, but the notion that nature is a perfectly ordered machine was an idea much more difficult to let go of.

Because of this providential theology, the existence of struggle, pain, and death in nature was frequently a difficult topic for Enlightenment-era naturalists. It is sometimes assumed that for this reason, prior to the nineteenth century, virtually no naturalists accepted the reality of extinction. Even in the later eighteenth century, some naturalists denied the possibility of extinction on effectively providential theological grounds. One of the most famous examples of such extinction denial is Thomas Jefferson's *Notes on the State of Virginia*, in which Jefferson considered the recent discovery of the fossil mastodon—or the "American incognitum," as it was sometimes called—a challenge to the stability of nature's economy. The problem, of course, was that this fossil appeared to represent an animal that had no living representatives. But Jefferson argued, as did some of his contemporaries, that living members of groups of apparently extinct animals like the incognitum simply had not been discovered yet. The North American continent was an enormous place, after all. For this reason, Jefferson included the "mammoth" in his list of extant North American species in *Notes on the State of Virginia*, and justified this decision by explaining, "Such is the economy of nature, that no instance can be produced, of her having permitted any one race of her animals to become extinct; of her having formed any link in her great work so weak as to be broken" (Jefferson 1785, 77).

In adopting this stance, Jefferson was in good company. Earlier in the century, the great savant Wilhelm Gottfried von Leibniz had also denied the possibility of extinction, and closer contemporaries, like the French naturalist Louis Jean Marie Daubenton, advanced anatomical arguments that the incognitum was within the normal variation of living pachyderms. However, despite entrenched cultural objections to the notion, a number of late-eighteenth-century naturalists did in fact regard extinction as a viable explanation for many of the fossil discoveries that were being unearthed in Europe and North America with increasing frequency. In fact, as the eminent historian of geology Martin Rudwick has convincingly demonstrated, by the end of the eighteenth century the central intellectual question surrounding extinction was not whether it had ever occurred, but rather how and with what frequency it had taken place, and how it was to be understood within broader emerging understandings of the historicity of the earth.[3]

Extinction and Catastrophe

The disciplinary locus for many of the most important debates about extinction over the last two centuries has been paleontology, since fossil specimens have provided the most obvious testimony to the vast numbers of organisms that have become extinct over the history of life. As this book will show, paleontologists have also made many of the most important theoretical contributions to the study of extinction, since extinction dynamics and patterns are often only interpretable at a resolution of tens or hundreds of millions of years. At the beginning of the nineteenth century, however, paleontology was in its infancy as a professional scientific discipline, and the interpretation of fossils was complicated by uncertainties as basic as the approximate age of the earth and the nature of historical geological processes. Early discussions of extinction, then, were bound up in broader debates about the earth's past and the tempo and mode of geological change. These uncertainties about the earth's geological past were compounded by equally vexing biological questions related to the possibility and nature of organic development, the fixity of species, the taxonomic organization of organisms, and the interpretation of what we would now call ecological relationships. It would take at least another century before most of these geological and biological problems were settled, but they contributed to a lively debate throughout the nineteenth century that occupied naturalists and "savants" across Europe.

One of the central problems in what Rudwick has called the "discovery of geohistory" centered on whether the earth's history has been characterized by a steady, gradual unfolding of geological processes, or has rather been "punctuated" by episodes of sudden and drastic change. The seventeenth-century natural philosopher Nicolas Steno is often credited with the discovery that the layers of the earth have not always existed in their current state and arrangement, and with the realization that the study of existing strata could unravel a history of the earth's past. Subsequent authors, such as the Scottish geologist James Hutton, the French naturalists Georges Cuvier and Alexandre Brongniart, and the English surveyor William Smith, expanded these principles as the basis for the modern science of stratigraphy, which by the 1820s allowed naturalists to map the order and locations of strata and to construct an approximate, relative geological timescale. Importantly, as Steno was among the first to recognize, characteristic fossils in each stratum could be used as a key to distinguish layers from one another.

By the late eighteenth century, Hutton and others recognized that profound geological changes had taken place in the earth's past: mountains had thrust themselves through the earth's crust, continents had been lifted up and subsided into the sea, volcanoes had deposited massive amounts of molten rock, and layers of the earth's crust had become twisted and bent far out of their original positions. Depending on the pace of their operation, changes like these could have had a profound impact on living creatures; but there were major disagreements over how, and how quickly, the geological processes that shaped the strata acted. Hutton favored a model of geologic change in which these processes happened very slowly, over nearly unimaginable amounts of time—a model now commonly referred to as "uniformitarianism." In many ways, Hutton was committed to this uniformitarian model because he felt it reflected the kind of stately deployment of natural laws best exemplified in a Newtonian, deistic worldview. The basic argument is that ordinary geological processes of the type observable around us today could, given enough time, produce drastic cumulative structural changes. Thanks to the influence of the great nineteenth-century Scottish geologist Charles Lyell, who will be discussed below, this principle of uniformitarianism has become a central pillar of modern geology.

But other late-eighteenth-century observers, such as the French naturalist François-Xavier Burtin and the German polymath Johann Friedrich Blumenbach, developed a model in which the earth's history had been marked by drastic, catastrophic "revolutions" that produced sudden geological and organic change. The idea that catastrophic events have taken place in the past was not new: since the seventeenth century, physico-theologists like Ray, Thomas Burnet, and William Whiston had attempted to explain scriptural events like the Noachian flood using naturalistic processes. But authors such as Burtin explicitly framed these revolutions in a nonscriptural context, and attempted to develop a naturalistic account in which they were part of the regular course of nature. As Burtin put it in in his 1789 essay "Révolutions generals," "The surface of the globe is but a series of documents that demonstrate a series of revolutions on this planet" (Burtin 1789, 200).[4] Significantly, Burtin and Blumenbach argued that these geological revolutions had been accompanied by massive extinctions in which whole floras and faunas were wiped out but were ultimately replaced, in some mysterious process, by new ones.

While some, like Jefferson, continued to deny the reality of extinction, for most naturalists at the turn of the nineteenth century the real debate concerned whether extinction was in a Huttonian sense a uniform and gradual process, happening only slowly or rarely, or rather a matter of catastrophic mass extermination. This debate obviously implicated basic ideas about the balance of nature and the dynamics of change. Given the available evidence, it seemed equally possible to describe geohistory either in terms of a balanced equilibrium of processes that evened out to produce a "steady state," or alternatively as a record of disequilibrium and catastrophe. These debates also invoked differing theological commitments; the balanced equilibrium of uniformitarianism sat more comfortably with those naturalists who subscribed to a deistic theology in which God acted on the universe through invariant natural laws, and literal interpretation of scriptural events was eschewed. "Catastrophism," on the other hand, often—but not always, as we will see—found favor with scientists who sought to explain particular historical events described in scripture, such as the Noachian flood, that appeared to require special explanations. This was certainly the case with some British catastrophists, including very prominently the British geologists Robert Jameson and William Buckland. The immediate context for these arguments was the interpretation of fossils and geological processes, but the broader stakes invoked strikingly different understandings of the tempo of historical change and the regularity of natural processes. The extinction imaginary of the early nineteenth century, then, hinged precisely on how the newly discovered empirical evidence of geology supported or impinged upon this set of broader theological and cultural beliefs.

The resolution of this debate hinged on fossils. Absent any absolute dating techniques (radiometric dating would not be available until the early twentieth century), the fossil record gave the best available evidence about the suddenness, magnitude, and generality of geologic change. If it could be determined that many fossil species had become extinct in a coordinated fashion at one or more points in the geological record, then this might lend support to theories proposing sudden revolutionary change. If, on the other hand, evidence of piecemeal extinction or even transmutation (evolution) predominated, then the more gradual uniformitarian model would seem to be favored.

The most important early-nineteenth century figure in this debate— and one of the most important nineteenth-century theorists about extinction—was Georges Cuvier. Born in 1769 to a bourgeois Protestant family in a French-speaking region of Germany which later became part of France, Cuvier combined early training in zoology with an interest in fossils and was appointed, while still only in his mid-twenties, to the newly established Musée national d'histoire naturelle in Paris, where he spent his entire career, eventually holding a professorship and a peerage in recognition of his stature. He was regarded during his lifetime as perhaps the most influential naturalist in France, if not Europe, and is recognized as having helped establish the study of comparative anatomy as an important scientific field. His anatomical reconstructions of fossil vertebrates, often based on only a few bones, are still considered brilliant, and he produced a broad theoretical revision of the Linnaean taxonomic system that was extremely influential in its day.

But Cuvier will always be most widely remembered—fairly or unfairly—for promoting a theory of earth's history in which the geology, flora, and fauna of the globe have periodically been radically altered by a series of catastrophic "revolutions." This theory was presented in a lengthy introduction ("Discours préliminaire") to his mammoth work on fossil vertebrates *Ossemens Fossiles* (1812), and subsequently revised and published on its own as *Discours sur les révolutions de la surface du globe* in 1826. It was also published in unauthorized and modified English translation in 1813 by the Scottish geologist Robert Jameson as

Essay on the Theory of the Earth, where it was read by many of the most prominent British naturalists of the time. Jameson's translation altered some aspects of the theory to give the distorted impression that Cuvier's work was an attempt to accommodate geology to scripture, which it was not; and it also mischaracterized some of Cuvier's views about the regularity of natural processes. In large part on the basis of Jameson's translation, Cuvier is now often remembered—and denigrated—as the father of a speculative and religiously motivated "catastrophism," which was vanquished by the proper, rational "uniformitarianism" of Charles Lyell. Though often repeated in textbooks and historical accounts, this characterization is a drastic oversimplification of a much more complex and interesting history.

Cuvier was undoubtedly one of the most important early proponents of biological extinction, and his views—presented in his own and other popular accounts and lectures—helped legitimize extinction not just among fellow naturalists, but to a wider educated public in Europe and North America. His brilliant reconstructions and interpretations of large fossil vertebrates—such as the mastodon or "American incognitum"—helped definitively establish that these were extinct creatures with no close living relatives. It was in the course of his studies of these large extinct vertebrates that Cuvier's more general theory of earth's history took shape; one fact that had prevented authors such as Jefferson from accepting extinction was the apparent well-adaptedness and robustness of the specimens being discovered. What, they wondered, could have caused the majestic mammoth, which appeared ideally suited for the American plains, to have died off? This troubled Cuvier as well; and as more large fossil vertebrate types were discovered, it led him towards the conclusion that only some kind of significant and widespread environmental catastrophe could have done the job.

It also encouraged Cuvier to pay close attention to differences in the strata in which fossil specimens were found. This ultimately led him back to the much earlier fossil invertebrates found in the geological strata around Paris, to which he gave a comprehensive survey with his colleague Alexandre Brongniart in the 1810s (fig. 1.1). Cuvier's conclusion was that a preponderance of evidence—the apparent adaptedness of fossil forms, the significant changes in geology at stratigraphic

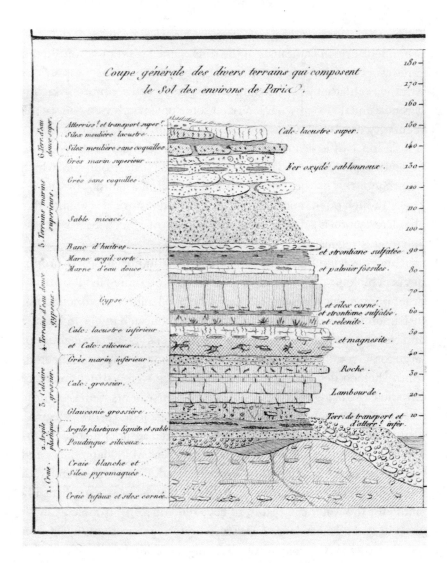

FIGURE 1.1 A classical stratigraphic visualization of ideal cross-sections of the earth's layers. From Georges Cuvier and Alexandre Brongniart, *Description géologique des environs de Paris* (Paris, 1812).

boundaries, and the sheer number of extinct taxa discovered—argued that

> life on earth has often been disturbed by terrible events: calamities which initially perhaps shook the entire crust of the earth to a great depth, but which have since become steadily less deep and less general. Living organisms without number have been the victims of these catastrophes. Some were destroyed by deluges, others were left dry when the seabed was suddenly raised; their races are even finished forever, and all they leave in the world is some debris that is hardly recognizable to the naturalist (Cuvier 1831).[5]

However, it is important to stress that Cuvier did not associate these revolutions with supernatural or scriptural events, nor did he necessarily believe that they required mechanisms outside of the ordinary run of geological processes. In this sense, his view challenges a simplistic dichotomy between deist uniformitarians and literalist catastrophists. For instance, he believed that the most recent revolution that wiped out the mastodon and other large vertebrates was most likely the result of an enormous tsunami, and he cited geological evidence for what he interpreted as the effects of a massive flood in the layers where the fossils were found. Admittedly, there is no precedent for a potentially continent-wide flood in recorded human history, but Cuvier was among many contemporary naturalists on whom the realization was dawning that human history was but a tiny sliver of the overall history of the earth. What made Cuvier's ideas potentially objectionable to contemporary naturalists of a more "uniformitarian" persuasion was not the kind, but rather the magnitude of the events required to produce widespread, even global, mass extinction. But Cuvier was hardly alone in speculating about catastrophic mass extinctions. His compatriot Élie de Beaumont promoted an account of periodic revolutionary catastrophe caused by "mega earthquakes," and in England William Buckland supported a "diluvial" theory of mass extinctions, to give just two prominent examples.

It is important to point out here that the term "catastrophism" has always been a remarkably flexible and imprecise term. Some eighteenth- and early-nineteenth-century geohistorical theories proposed some kind of extraordinary, perhaps unprecedented mechanism — a "catastrophe" — as the trigger for major transformations of the earth and its inhabitants. As we have seen, Cuvier's theory only partially fits this description, because though his proposed mechanisms were indeed dramatic, they were extrapolated from known natural phenomena — unlike, for example, William Whiston's early-eighteenth-century hypothesis that a passing comet's tail deposited the waters documented in the flood story in Genesis. Another way catastrophism might be construed is by postulating that "revolutions" of some kind have taken place either singularly or with some periodicity in the earth's past. But while today we associate the term with a sudden, violent upheaval, in the contemporary context a "revolution" could be applied to any significant change, whether or not that change was sudden or violent. Indeed, Burtin, Jean-André de Luc, Constant Prévost, Brongniart, Cuvier, and even Lyell in his early writings all acknowledged that "revolutions" could be gradual as well as sudden affairs, and need not in fact be "catastrophic" at all. Catastrophism has often been taken to imply that geological periods have been separated by fairly distinct environmental or faunal changes. But a simple empirical fact apparent to any geologist involved in the reconstruction of stratigraphy is that individual strata can be identified precisely *because* they contain obvious and significant differences in geological and faunal composition. Indeed, stratigraphy itself is founded upon observations of sharp breaks in the type of rock and the kinds of fossils found from one layer to the next, which are used to define geological periods themselves. The distinctness of geological strata is a generally agreed-upon fact that is, in principle, agnostic toward a broader theoretical interpretation of what those differences mean. In both contemporary and historical (retrospective) accounts, catastrophism was often associated with a young-earth biblical chronology, giving it for naturalists of Darwin's generation and beyond more than a whiff of the supernatural. While it is true that some, like Buckland, associated geological catastrophes with the Noachian flood, this was actually a minority position, and it in no way reflected Cuvier's interpretation, which was thoroughly naturalistic.

Finally, catastrophism is often associated with some theory of mass extinction—the idea that a large number of species died out in a coordinated fashion within a relatively short period of time, or were extinguished because of some individual event. But again, mass extinctions do not require sudden or violent causes. In the 1820s, Prévost interpreted the succession of Tertiary environments and organisms around Paris that Cuvier and Brongniart had described as a much more gradual process than Cuvier had proposed. Likewise, Cuvier's sometime collaborator Brongniart hypothesized that extinctions were caused by temperature changes as the earth slowly cooled over time. And other naturalists, like John Fleming, were beginning to suspect that even mass extinctions need not be all-or-nothing affairs. In some cases, groups of species in a fairly limited geographical region might indeed have become extinct at roughly the same time; but if extinction is a pervasive phenomenon, then sudden worldwide events are not required to explain even the apparently dramatic faunal turnover exhibited in the fossil record.

The point here is that the term "catastrophism" since its very first application in the debates of the 1830s, has been something of a straw man. It is extraordinarily flexible, and at the same time remarkably imprecise—it could be applied either to nearly everybody or to nobody at all. Like many straw men, however, it is tremendously important for what it says about the attitudes of those who deployed it—in this case, almost exclusively as an epithet by supporters of Lyell's "uniformitarian" geological theory and their intellectual descendants—and for how it has influenced and often constrained scientific discussion over the past two centuries. It is also useful, for heuristic purposes, to use the terms "catastrophist" and "uniformitarian" when describing debates and battle lines as they were understood at the time. I am certainly not claiming that there were no substantive disagreements about the causes, magnitude, or consequences of extinction among the scientists I am discussing. Rather, my point is that these terms—"catastrophism" and "uniformitarianism"—are actors' categories that invoke those enormously significant disagreements which are the subject of this book, and that their complex scientific and cultural history cannot be reduced to the binary opposition of labels.

Extinction and Internalism in the
Early Nineteenth Century

Even if the catastrophist/uniformitarian dichotomy is not the most illu-minating perspective with which to interpret this history, there were nonetheless other legitimate barriers to consensus about the nature of extinction during the first half of the nineteenth century and beyond. I would argue that a far more important distinction in the history of ideas about extinction relates to whether the causes of extinction were under-stood to be *extrinsic*—that is, caused by some environmental change or other factor operating on populations of organisms—or rather *intrin-sic*—in other words, caused by factors internal to the organisms them-selves or to the dynamics of their populations. This dichotomy maps approximately, though not universally, to beliefs about the possible magnitude of extinctions. A theory that posits mega-tsunamis periodi-cally sweeping across continents and wiping out hundreds of species in an instant is a theory of *mass* extinction triggered by extrinsic forces. A theory that extinction results from the "racial senility" of individual species (which we will see an example of shortly) is a theory of piece-meal, intrinsic extinction. There are certainly many examples of theo-ries that are somewhat less neatly categorizable in this way: for ex-ample, Brongniart's suggestion that slow climate change caused gradual extinctions, or Fleming's hypothesis that extinction of megafauna like the mammoth was caused by human predation (hunting). Nonetheless, one can broadly claim that that nearly all theories of mass extinction hy-pothesize external mechanisms while many, though not all, theories of gradual or piecemeal extinction tend to assume internal causal factors.

The internal/external dichotomy could also condition whether or not a naturalist accepted extinction as a natural phenomenon at all. Probably the most famous nineteenth-century example of this was Jean-Baptiste Lamarck's position. Lamarck, a contemporary colleague of Cuvier at the Musée national d'histoire naturelle, is famous for having presented the first systematic theory of organic transmutation or evo-lution. This theory was different from Darwin's eventual theory of de-scent with modification by natural selection in two important respects. In the first place, Lamarck believed that characteristics acquired by an organism during its lifetime could be passed on to its offspring; (for example, a giraffe that craned its neck to reach leaves at the top branches of trees might pass along a slightly longer neck to its offspring. Although Darwin himself considered the inheritance of acquired characteristics as a potential mechanism for transmitting some physical and behavioral characteristics, this mechanism was mostly rejected in the eventual received view of "Darwinism." Secondly, and more important, Lamarck believed that evolution—that is, the transmutation of one species into another—took place along a preordained pathway of lesser to greater complexity, and was guided by an *internal* mechanism, which he termed "the power of life" and likened to the force of gravity or other "imponderable fluids" like electricity or magnetism. While the nineteenth and early twentieth centuries saw a number of similar "internalist" evolutionary hypotheses (such as orthogenesis, or directional evolution), Darwin staunchly rejected all such directional or internal mechanisms.

When in the early nineteenth century it became impossible to deny that many fossil types had no analogous living representatives, really only two alternative explanations presented themselves: either many once-living species had become extinct, perhaps to be replaced somehow by new and different ones, or else those species had changed over time to become the species we see around us today. It is often assumed that evolution was widely rejected prior to Darwin because it was too "radical" for the worldview of nineteenth-century naturalists. In fact, permanent extinction was a potentially far more radical alternative, implying as it did that nature might not always maintain a stable equilibrium. Lamarck certainly found this to be the case; for most of his career he avoided any transmutationist thinking, and it was only when confronted by the mounting evidence from fossils being accumulated by people like Cuvier that he abruptly converted, in large part because of his deep commitment to a balance of nature that he saw threatened by the specter of permanent extinction. Lamarck's evolutionary theory, then, was explicitly motivated by notions of nature's economy that were very similar to those expressed by Linnaeus fifty years earlier. The mechanism that governed transmutation was, he stressed, an intrinsic and perhaps divinely inspired natural law: "Nature (or her Author) in creating animals, foresaw all the possible kinds of environment in which

they would have to live, and endowed each species with a fixed organisation and with a definite and invariable shape, which compel each
species to live in the places and climates where we actually find them,
and there to maintain the habits which we know in them" (Lamarck
1809).[6]

Lamarck's vision of nature is thus as a dynamic equilibrium: environments change very slowly over time—this can be observed even on
the scale of a single human lifetime—and nature has provided a natural
process that ensures that organisms remain adapted to their stations. To
Lamarck, permanent extinction implied a failure of nature's economy.
As he put it, the problem of extinction involved asking whether "the
means which nature adopted to assure the conservation of species or
races has been so inadequate that entire races have now been wiped out
or lost" (Lamarck 1809).[7] While he granted that the deliberate extermination of species by humans was "a possibility," he argued that most organisms—especially those that lived in the seas—"are protected against
the destruction of their species at the hand of man." Since most apparently extinct species in the fossil record are marine bivalves, human
agency cannot be blamed for their disappearance, thus leaving only two
possibilities: either they do have living representatives that simply have
not yet been discovered, or else those earlier forms have transmuted
into something different. The surprising fact, Lamarck argued, is not
that we find so *few* fossils with living analogs, but rather that, given the
ubiquitous action of transmutation, we find *any* with living representatives.

Lamarck then went on to complain that those "naturalists who have
not perceived the changes which most animals experience with the passage of time . . . have assumed that a universal catastrophe took place
with respect to the terrestrial globe and destroyed a large number of the
species then in existence." While he granted that natural phenomena
like earthquakes and floods could cause localized disorder, he denied
the need to invoke catastrophic agents, and expressed his "pity that this
convenient method of dealing with one's embarrassment when one
wants to explain the operations of nature whose causes one been unable to grasp [*sic*] has no foundation except in the imagination which
created it" (Lamarck 1809). Ultimately, then, Lamarck considered mass

extinction not only unnecessary but even offensive to the dignity of nature. As he concluded,

> If one considers, on the one hand, that in everything which nature brings about, she makes nothing abruptly and everywhere works slowly and by successive degrees and, on the other hand, that the particular or local causes of disorders, revolutions, displacements, and so on, can provide reasons for everything which we observe on the surface of the earth and are nonetheless subject to nature's laws and her general progress, one will recognize that it is not at all necessary to assume that a universal catastrophe came to knock over everything and destroy a large part of the very operations of nature (Lamarck 1809).

In regard to the power of life, we might say that Lamarck's internalism was so extreme that it led him to deny the possibility of extinction at all; but there were other ways that extinction could be made compatible with an internalist philosophy of biology. One of the most influential though now largely forgotten early theorists of extinction was the Italian naturalist Giambattista Brocchi, whose 1814 treatise *Subapennine Fossil Conchology* was read and admired by geologists throughout Europe, including Lyell and eventually Darwin. The work itself was a survey of fossil mollusks found in Italian deposits dating from what contemporary geologists referred to as the Tertiary period, or the "third age" of earth's history. While we now date those rocks to between 2.6 and 65 million years old, the influential eighteenth-century geochronology of Giovanni Arduino located the Tertiary as contemporaneous with the Noachian flood and other events of Genesis. However, by the early nineteenth century many naturalists believed that Tertiary strata were far older, perhaps having been deposited tens of thousands of years before the earliest recorded human history. Because these deposits, which were understood to hold the earliest record of life on earth, had great significance for the broader reconstruction of geohistory, the establishment of a relative dating and stratigraphy for the Tertiary was an important geological problem at this time, and Brocchi regarded his work as "a series of documents that shed light on the ancient history of the globe" (Brocchi 1814).[8]

Growing up in the foothills of the northern Italian Alps, Brocchi took an interest in fossils from an early age, and by the time he began serious study of geology in his early thirties he was well aware of wider European debates about the extinction of species. His own fossil collection and research was leading him to the same general conclusions that Cuvier and others were reaching at the same time: that many of the fossils being discovered appeared to have no living analogs. However, whereas Cuvier explained the problem with a theory of catastrophic mass extinction, and Lamarck with transmutation, Brocchi adopted a solution that was for its time very unconventional. Noting that many fossil deposits contain a mixture of apparently extinct and extant species, Brocchi began developing the idea that species become extinct in a piecemeal fashion, not because of catastrophes or external mechanisms, but rather because species, like individuals, have natural "life spans." As early as his 1807 *Mineralogical Treatise*, Brocchi argued that it is "a constant and general law of Nature" that "species die just like individuals do," because of "the lack of reproductive force and the inability to develop" (Brocchi 1807).[9]

This was, to say the least, a rather unorthodox position for a naturalist to take in the early nineteenth century. Brocchi was proposing, in effect, that species have "births" and "deaths" just as individual organisms do, and that the cause was entirely *natural*, produced by "a gradual and constant law" of nature (Brocchi 1807).[10] This was an explanation that rejected scriptural geological accounts (such as Buckland's) that were still popular, and which also obviated the need for great catastrophic revolutions to explain anomalous fossil organisms. It should be emphasized that, while Brocchi had no clear mechanistic account of the nature of the force that created species or determined their longevity, his position was guided by empirical considerations: *pace* Cuvier, he simply did not see evidence that species had become extinct en masse as the result of a single event. But it is also clear that in many ways Brocchi's position was influenced by a conservative sensibility towards the economy of nature. While geological evidence would not allow him to conclude that nature conserved each species eternally, he was nonetheless committed to a view of nature as a balanced, if dynamic, equilibrium. For Brocchi, change was just as "natural" as permanence, so he argued:

> Why don't we thus admit that species die like individuals, and that like them they have a fixed and determined period for their existence? This should not seem awkward, if we think that nothing is in a state of permanence on our globe, and that Nature is maintained actively with a perpetual circle and a perennial cycle of changes (Brocchi 1807).

In terms of its consequence for traditional ideas about the balance of nature, this stance is actually significantly less radical than Cuvier's, which proposed a directional, nonequilibrium model of geohistory in which sudden transformations periodically take place that radically alter the earth and its inhabitants. Cuvier's view could be interpreted to deny both the beneficence of a divine creator and a Newtonian clockwork regularity to nature's operations. In contrast, in Brocchi's theory change is conserved: as one species dies, another is born to take its place, and the "cycle" continues. Brocchi also emphasized that this process takes place very slowly, since "by imperceptible grades species come to their annihilation"; and he noted that many species appear to have persisted for very long periods of time. Nor did Brocchi's internalist theory imply transmutation or evolution; one of his central observations was that species that persist for long periods of time do not appear to change. It did require a perhaps uncomfortable acknowledgment of the prevalence of death and destruction in nature, implying that "Nature in some way more likely pleases herself in degrading and destroying her works, than in perfecting them and extending their conservation." But Linnaeus had already shown that death could be conceived as an essential part of nature's economy, and Brocchi's theory was far less violent and arbitrary than Cuvier's, imagining extinction as the rather peaceful conclusion to the natural life span of a species, rather than the terrifying result of a horrific catastrophe.

Brocchi's theory then was in many ways a very clever and somewhat radical way of accounting for overwhelming evidence that extinction was a natural and even common phenomenon, while maintaining older cherished beliefs about the economy of nature and the regularity of natural laws. Although it did not achieve a great deal of notoriety, it was quietly influential. Ultimately, naturalists like Lyell and Darwin emphasized the role of the environment much more than intrinsic factors in the extinction and development or evolution of individual organisms and species. But both authors found occasion to refer favorably to Brocchi, and Brocchi's theory is an important link in a lineage of naturalistic thought that understood extinction as a gradual and inevitable process that contributed to a balanced economy of nature. And as we will see in later chapters of this book, Brocchi's analogy—the notion that species could be conceived in many evolutionary and ecological respects as individuals—had an important resurgence during the 1970s and 1980s, in the context of debates surrounding the interpretation of patterns of diversification and mass extinction.[11]

Extinction, Uniformity, and the Balance of Nature: Charles Lyell

We see many of the currents of the contemporary debate surrounding extinction—intrinsic versus extrinsic causes, piecemeal versus mass extinction, gradual versus sudden operation, balance versus disequilibrium—come together in a powerful and influential interpretation in the work of the Scottish geologist Charles Lyell. Lyell, who trained as a barrister but spent his career advancing the theoretical development and professionalization of geology, is by general acknowledgement one of the central figures in the history of nineteenth-century British science. His ideas had a deep influence on Darwin, and the two naturalists became close friends and correspondents, exchanging hundreds of letters over several decades between the 1840s and the 1870s. Even more broadly, Lyell's ideas about natural change and balance profoundly influenced scientific understanding of the nature of geological and organic change by viewing these processes as components of a linked, natural equilibrium. Lyell's view of extinction ultimately hinged on the dynamic relationship between organisms and their slowly changing environments, a notion which took on even greater resonance—as we will

explore in the next chapter—in the context of Darwin's emerging evolutionary ideas. Finally, by explicitly linking extinction with processes
and patterns in the history of life's diversity, Lyell helped install extinction as a central component in scientific interpretations of the patterns
and processes that have shaped the diversity of life on earth.

In many accounts of the history of scientific attitudes towards extinction, Lyell is presented and often celebrated as the man who definitively put paid to Cuvier's "catastrophist" theory of revolutionary
mass extinction. While this version of history has achieved a kind of
mythological status through constant repetition in geology and paleontology textbooks, it was also in many ways a caricature of Lyell's own
devising, given that Lyell himself was largely responsible for promoting
the distinction between uniformitarianism and catastrophism in his
own writings. While it is true that Lyell was sharply critical of Cuvier
in his monumental *Principles of Geology* (1830–33), the reality is somewhat more complex. As a young man looking to make a name for himself in British scientific circles, Lyell was a frequent contributor to the
Tory magazine *Quarterly Review*, which gave him an influential voice
among an elite readership. One essay in particular—an 1826 review of
an annual volume of the *Transactions of the Geological Society*—dealt
directly with the problem of extinction. Here Lyell cited, "among facts
and conclusions now universally conceded," the conclusion that geological strata "have been subject, at different, and often distant, epochs,
to violent convulsions" (Lyell 1826, 507). This essay was generally quite
favorable toward Cuvier's interpretations of extinction, supporting the
French naturalist's conclusion that many fossil species are genuinely
extinct, as opposed to having undiscovered living representatives, and
that those animals belonged to "an earlier epoch . . . peopled with a race
of terrestrial quadrupeds of an entirely different description; a race, of
which most of the genera and all the species known to us in fossil remains have since been annihilated" (Lyell 1826, 511).

As for the causes of this "annihilation," Lyell freely speculated that
floods or earthquakes could have been the culprit, and he even acknowledged that Cuvier's opinion that observable phenomena were not
sufficient to explain these extinctions was "entitled without doubt to the
more respect." However, Lyell also cautioned that it was "premature

to assume that existing agents could not, in the lapse of ages, produce such effects as fall principally under the examination of the geologist," noting that while there were "proofs of occasional convulsions . . . there are also proofs of intervening periods of order and tranquility" (Lyell 1826, 518). On balance, though, this essay is hardly the stinging rejection of Cuvier's revolutionary interpretation of earth's history that one might expect from the great "uniformitarian." Lyell seems to have been endorsing a view in which sudden "catastrophic" change was at least partially responsible for past extinctions, testimony for which he found in "the frequent unconformability of strata [which] clearly shows that disturbances have taken place at many and at different periods."

At the same time, Lyell appears to have been concerned with interpreting geological evidence in a framework that was at least broadly conformable with an orderly and perhaps divinely inspired natural economy. He may well have been inspired in his particular formulation of economy by the political economies of writers like Adam Smith, a fellow Scot with a deistic leaning similar to Lyell's own. Lyell was struck just as much by the appearance of new forms in the fossil record as by the disappearance of older ones, even though the mechanism by which new species are created remained mysterious. Noting that "successive races of distinct plants and animals have inhabited the earth," he argued that this was "a phenomenon perhaps not more unaccountable than one with which we are familiar, that successive generations of living species perish, some after a brief existence of a few hours, others after a protracted life of many centuries" (Lyell 1826, 538). This analogy between species and individual organisms would feature importantly in Lyell's later writings, and it was almost certainly influenced by Brocchi, whose ideas Lyell had encountered during his own study of European Tertiary formations. Ultimately, though not developed much further in his 1826 essay, this shows that even at a stage when he was unwilling to dismiss Cuverian revolutions, Lyell was drawn towards a naturalistic causal explanation for extinctions, one that did not rely on sudden catastrophes. In concluding his discussion of extinction, he argued that "sources of apparent derangement in the system appear, when their operation throughout a series of ages is brought into one view, to have produced a great preponderance of good; and to be governed by fixed general laws, conducive, perhaps essential, to the preservation of the habitable state of the globe" (Lyell 1826, 539).

This argument—that even the appearance of disorder or upheaval in the geological record does not necessarily disturb the underlying rational economy of nature—would be central to Lyell's continued theorizing about extinction. While Lyell was certainly the most important contemporary proponent of this view, it was not an unprecedented idea. For example, writing in 1804, the English physician and geologist James Parkinson (the first identifier of Parkinson's disease) argued that accepting the fact of extinction need not disturb the belief that nature exists in a balance, since "that plan, which prevents the failure of a genus, or species, from disturbing the general arrangement, and oeconomy of the system, must manifest as great a display of wisdom and power, as could any fancied chain of beings, in which the loss of a single link would prove the destruction of the whole" (Parkinson 1804, 468). In other words, Parkinson argued, nature could maintain an equilibrium even in the face of the extinction of the occasional species, or even genus, provided that the loss was made good with the creation of a new species somewhere else, thus preserving the divine rationality of "those laws, by which the regulation of the oeconomy of creation was decreed."

But, as Lyell's views developed, he increasingly shied away from any sense of directionality or irreversibility in the history of geologic and organic change; and at the same time he became more committed to slow, uniform physical processes as the source of that change. In part, this was the result of his growing firsthand knowledge of European Tertiary geological formations, which he experienced during travels in the late 1820s. This experience helped convince him that these geological deposits testified to an era that was both very ancient (much older than the few thousand years assumed by proponents of scriptural geology) and remarkably similar to our own in terms of environment and organismal diversity. In an attempt to establish a reliable relative chronology of these formations, Lyell collaborated with the French mollusk expert Paul Deshayes on an exhaustive "census" of more than three thousand Tertiary fossils, which he subjected to basic statistical analysis in order to determine the percentage of fossil organisms in each stratum that had become extinct. While this general project—which Rudwick describes as an attempt to create a "fossil chronometer" for the entire fossil record—was less successful than Lyell had hoped, it did provide him with one crucial insight: Just as Brocchi had noted, the invertebrate fossil record appeared to exhibit virtually no evidence of any catastrophic sudden mass extinctions. Rather, formations tended to contain a mixture of extinct and extant forms, suggesting that extinctions happened in piecemeal fashion, and that species generally appeared to persist for very long periods of time, thus suggesting that extinction is a gradual process. Furthermore, even extinct fossil mollusks do not appear to be radically different from their living relatives, and there are no examples of the radical differences in fauna that Cuvier discovered in more recent extinct American vertebrate remains. This fact suggested to Lyell that the environment of the distant past was quite similar to our own today, further casting doubt on the notion of successive and radically different global geological epochs. This view also sharply contrasted with more traditionally theistic interpretations of earth's history promoted by contemporary English geologists like Buckland and Adam Sedgwick, which tended to see the history of the globe as a succession of distinct environments punctuated by great geological catastrophes.

These observations contributed to a very different picture of extinction than Lyell had proposed in his 1826 essay, and this change is reflected in his magnum opus, *Principles of Geology*. Here Lyell developed the view that the earth's history was one of slow, cyclical environmental change, requiring the gradual adaptation of organisms to these changing environmental circumstances. Where organisms were unable to adapt—and Lyell allowed for limited organic modification, though not genuine transmutation—populations were required to either migrate to more hospitable locations or face inevitable extinction. Importantly, this was a process that was constantly ongoing; environmental changes, such as the rising and lowering of global sea levels or changes in temperature, were not caused by catastrophes or geologic revolutions, but were gradual, and their cumulative effects could only be detected on the order of many thousands of years. Finally, environmental change—and adaptation—was circumscribed by fairly limited boundaries, and it fluctuated back and forth along lengthy cycles. In this regard, Lyell's geology was "uniformitarian" in the same sense as that of James Hutton, who had concluded that "we find no vestige of a beginning, no prospect of an end" to the history of the earth. Lyell's key innovation, which made his work so important for Darwin, was that he linked the history of life to these processes of gradual geological change. In this way, the earth and its inhabitants existed in a perpetual state of dynamic equilibrium. A change in environment necessitated a corresponding adaptive response, but since environmental change was cyclical, there could be no ultimate direction to life's history.

Lyell treated the subject of extinction most directly in the second of the three volumes of *Principles*, which was published in 1832 (and was read avidly by Darwin, who was well into his five-year voyage on HMS *Beagle*). Here Brocchi remained an important influence, and Lyell explicitly endorsed the analogy between species and individual organisms, writing that the Italian "does not appear to have been far wrong" in his assertion that "the death . . . of a species might depend, like that of individuals, on certain peculiarities of constitution conferred upon them at their birth" (Lyell 1830–33, II:128). He also applauded Brocchi for rejecting catastrophic revolutions as the mechanism of extinction, and for instead "endeavor[ing] to imagine some regular and constant law by which species might be made to disappear from the earth gradually and in succession." However, Lyell rejected Brocchi's intrinsic mechanism of natural species life spans in favor of an external, environmental (one might even anachronistically say "ecological") explanation: "If it can be shown that the stations [i.e., "niches"] can become essentially modified by the influence of known causes, it will follow that species, as well as individuals, are mortal" (Lyell 1830–33, II:130).

Lyell also explicitly disavowed Cuvier's grand model of geologic change and resulting mass extinctions, although he did allow for periods of limited elevated extinction. "We are not about to advocate the doctrine of general catastrophes recurring at certain intervals," he wrote, nonetheless noting evidence of "important revolutions" that were "attended to by the local annihilation of many species. . . . without producing any extensive alterations in the habitable surface" (Lyell 1830–33, II:161–65). He stressed that these "revolutions" were nonetheless part of a balanced natural economy, and did "afford evidence in favour of the uniformity of the system, unless, indeed, we are precluded from speaking of *uniformity* when we characterize a principle of endless variation" (Lyell 1830–33, II:157). What this meant, in practice, was that if "we admit incessant fluctuations in the physical geography, we must, at the same time, concede the successive extinction of terrestrial and aquatic species to be part of the economy of our system" (Lyell 1830–33, II:168). In other words, Lyell concluded that constant, slow variation is itself a kind of uniformity or equilibrium, and that the "Author of Nature" had "ordained that the fluctuations of the animate and inanimate creation should be in perfect harmony with each other" (Lyell 1830–33, II: 159).

For all of its appearance as a gradual and natural process, Lyell frequently used violent terminology such as "war," "strife," "annihilation," and "destruction" to characterize this dynamic organic equilibrium. Organisms are in constant competition with their changing environments, and also with one another—as Lyell illustrated with a lengthy discussion of the "continual strife" between plant species—contributing to the dynamic vision of Lyell's model. Although change occurs very slowly, nature never stands still. In this view of nature, extinction is not just common; it is, for some species, inevitable. Since "species are subject to incessant vicissitudes . . . it will follow that the successive destruction of species must now be part of the regular and constant order of Nature" (Lyell 1830–33, II:141). Here Lyell's interpretation of extinction touched directly on the balance of natural diversity, which he conceived as a constant—though constantly fluctuating—equilibrium. There are a limited number of places or "stations" available for organisms to occupy, and as one species vacates its place, another must come along to occupy it. As he put it, "The addition of any new species, or the *permanent* numerical increase of one previously established, must always be attended either by the local extermination or the numerical decrease of some other species" (Lyell 1830–33, II:142). If this is not quite "nature red in tooth and claw," it is nonetheless a vision of the natural order in which competition plays a prominent role. At the same time, there is an overarching balance and harmony, since competition itself—and the inevitable death and extinction that follows from it—is a mechanism for preserving a dynamic natural equilibrium.

It is also worth emphasizing how committed Lyell was to the notion that balance requires a continual replacement of one species for another. He was purposely elusive about whether the process of new creation required a divine hand or natural causes, but it was easy for many readers to read some of the old providential natural theology into his system. This position connected directly with one of his major geological arguments: that geological processes had no directionality, and that formations (and hence environments) found in one geological age would inevitably return as climate oscillated slowly between the boundaries of the steady state. Infamously, this led Lyell to speculate that organic history had no directionality, implying that even long-extinct creatures might "return" when environmental conditions were favorable. As he wrote in volume 1 of *Principles*,

> We might expect, therefore, in the summer of the "great year" which we are now considering [i.e., the grand geological cycle of climate], that there would be a great predominance of tree-ferns and plants allied to palms and arborescent grasses in the isles of the wide ocean. . . . Then might those genera of animals return, of which the memorials are preserved in the ancient rocks of our continents. The huge iguanodon might reappear in the woods, and the ichthyosaur in the sea, while the pterodactyle might flit again through umbrageous groves of tree-ferns (Lyell 1830–33, II:123).

Needless to say, this comment occasioned no small surprise and even derision from Lyell's contemporaries. Despite the fact that Lyell had mentioned that it was possible that extinct higher taxa (i.e., genera) might return, and not specific species, his contemporary and geological opponent Henry de la Beche mocked this passage with a cartoon he reproduced for his friends, in which Lyell appeared in the character of "Professor Ichthyosaurus," lecturing to a group of saurians on the topic of the past extinction of the human species (fig. 1.2). De la Beche's mockery aside, however, this incident reinforces the centrality of cyclical change and replacement in Lyell's vision of the economy of nature.

In presenting this view of nature, Lyell was not above drawing conclusions relevant to his own contemporary society. Lyell considered

FIGURE 1.2 A cartoon drawn by the English geologist Henry de la Beche depicting Charles Lyell as "Professor Ichthyosaurus," lecturing a group of students about a fossil human skull. The caption reads: "'You will at once perceive,' continued Professor Ichthyosaurus, 'that the skull before us belonged to some of the lower order of animals; the teeth are very insignificant, the power of the jaws trifling, and altogether it seems wonderful how the creature could have procured food.'" Lithograph by Sir Henry de la Bèche (1830), after his drawing. Credit: Wellcome Collection. CC BY.

humans as a potential agent of extinction, both in the distant past (as perhaps the cause of the extinctions of the megafauna Cuvier had reconstructed) and in more recent times, as in the famous examples of the dodo and the moa. He never doubted that human beings could cause the extinction of species, since "man is, in truth, continually striving to diminish the natural diversity of the *stations* of animals and plants, in every country, and to reduce them all to a number fitted for species of economical use" (Lyell 1830–33, II:147–48). He considered this to be an inevitable result of European imperial expansion, and argued, "We must at once be convinced, that the annihilation of species has already been effected, and will continue to go on hereafter, in certain regions, in a still more rapid ratio, as the colonies of highly-civilized nations spread themselves over unoccupied lands" (Lyell 1830–33, II:156). Yet he saw this as little cause for regret, arguing that "if we wield the sword of extermination as we advance, we have no reason to repine at the havoc committed, nor to fancy, with the Scottish poet, that 'we violate the social union of nature.'"[12] Why? Because extinction is part of the natural order of nature:

> We have only to reflect, that in thus obtaining possession of the earth by conquest, and defending our acquisitions by force, we exercise no exclusive prerogative. Every species which has spread itself from a small point over a wide area, must, in like manner, have marked its progress, by the diminution, or the entire extirpation, of some other, and must maintain its ground by a successful struggle against the encroachments of other plants and animals (Lyell 1830–33, II:156).

Furthermore, in language that would strike most modern readers as callous at the very least, Lyell made it clear that this explanation applied equally to the extinction of "races" of human beings: "A faint image of the certain doom of a species less fitted to struggle with some new condition in a region which it previously inhabited, and where it has to contend with a more vigorous species, is presented by the extirpation of savage tribes of men by the advancing colony of some civilized nation" (Lyell 1830–33, II:175). For this he offered no apology since, as he was quick to note, he viewed this as the natural and *inevitable* course of nature: "Few future events are more certain than the speedy extermination of the Indians of North America and the savages of New Holland in the course of a few centuries, when these tribes will be remembered only in poetry and tradition."

There is, therefore, little sense that Lyell was concerned that extinction, whether caused by environmental change or by interspecies competition, was a threat to the balance of nature. There is certainly no evidence that he believed that natural diversity needed to be protected, or that human beings should actively combat extinction. This is not to say that he did not appreciate or value the diversity of living things; he simply assumed that a diversity of life would be guaranteed as an inevitable consequence of the equilibrium of natural processes. As we will see in the next chapter, Lyell's vision of a dynamic but ultimately balanced equilibrium, in which extinction was the inevitable consequence of change and competition, and where the economy of nature was maintained by continual replacement, was largely adopted by Darwin. It was, as I will argue, part of the scientific and cultural foundation of the age in which both men lived, and central to the Victorian extinction imaginary. Extrapolations of lessons about extinction from the nonhuman biological world were frequently made to the context of contemporary European society, and biology was often used as justification for political expansion. It is just as much the case that biological ideas—about competition, and about the inevitability of failure and extinction—were influenced by existing social views. By the 1840s, extinction had been naturalized; but it was also inextricably bound up in cultural and political values about race, progress, and diversity. And by virtue of the fact that these scientific ideas about extinction, diversity, and the balance of nature—incorporated as they were into the foundation both of Darwinism and the emerging science of ecology—became so influential, the Victorian context and the values they represented continued to have influence long after they dropped from explicit view.

2

EXTINCTION IN A VICTORIAN KEY

During the 1830s, the British Parliament convened a series of hearings to consider troubling reports that had begun to filter back from its colonial outposts in South Africa and Australia about the relationship between British colonists and the native inhabitants of those lands. While a central objective of the imperial enterprise was, of course, to simply appropriate as much land and natural resources as possible, there were practical and humanitarian factors to be considered as well. From a cynical point of view, native peoples were vital to imperial expansion as a cheap source of labor, meaning that they could not simply be exterminated without consequence. And from a moral perspective, a significant element of the rhetoric surrounding empire was that imperialism was a divinely sanctioned, and perhaps natural, imperative to bring "civilization" to the benighted peoples of the globe.

Accordingly, in an 1831 parliamentary report produced from these hearings, correspondence between British Colonial Secretary George Murray and George Arthur, lieutenant governor of the penal colony at Van Diemen's Land (Tasmania), was entered as evidence of significant potential problems. In letters to his colonial representative, Murray expressed concerns about reports of the "great decrease" that had recently taken place in the local aboriginal population. Noting that it was "not unreasonable to apprehend that the whole race of these people may, at no distant period, become extinct," Murray concluded that "it is impossible not to contemplate such a result of our occupation of the island as one very difficult to be reconciled with feelings of humanity," and observed that "the extinction of the Native race, could not fail to leave an indelible stain upon the character of the British Government" (Sir George Murray to George Arthur, November 5, 1830).[1] Similar qualms were expressed in an 1835 report on conditions of native peoples in the Cape of Good Hope, where a Mr. Collins decried the "indiscriminate massacre" of local populations by Dutch settlers—"The total extinction of the Bosjeman race [Bushmen] is actually stated to have been at one time confidently hoped for"—but reported that fortunately the recent intervention of British authorities had prevented this. In a separate letter from the same report, a Mr. Moodie similarly applauded British intervention in Dutch massacres of the "Caffres." Although he insisted that it was not his place to determine "whether it is an inevitable provision of nature that the weaker must in one way or another melt away before the stronger power," Moodie nonetheless described "the success of the attempt to depart from the usual course, and to preserve the character and independence of the savage *after* he has been permitted to become acquainted with the possessions of his improved neighbor" (British parliamentary papers 1835, 40, 175).[2]

But a certain degree of fatalism was also present in many of these parliamentary reports as well. In 1835, a Select Committee on Aborigines was convened for hearings to determine future policy with respect to natives living in areas of British colonization. The cooperation of the Anglican church in this program was vital to the government's political objectives, since missionaries were on the front lines of the "civilizing mission," and church representatives held positions of authority in most major colonial centers. One of the witnesses before the committee was the head ecclesiastical representative in New South Wales, Archdeacon Broughton, who testified to the depressing nature of his experiences with the Aborigines he encountered. Part of his mission was to "civilize" the natives, and Broughton reported that these efforts appeared to be "hopeless"—not so much because the natives were unintelligent, but rather because they were so "entirely abandoned" to "ignorance and degradation" that the "expense" of the effort was not worthwhile.[3] Ultimately, he predicted that wherever Europeans meet with them, they appear to wear out, and gradually to decay; they diminish in numbers. . . . The tribe is gradually reduced from its original number to a much smaller number; it is a continual process of decay I should think, and it leads me to apprehend, that within a very limited period, those who are very much in contact with Europeans will be utterly extinct; I will not say exterminated, but they will be extinct (Broughton 1836, 17).

In other words, Broughton argued, "decay" and extinction were the *inevitable* results of contact with European settlers. This occurred not, as in the case of Dutch massacres, because of a deliberate program of extermination, but rather as the natural outcome of cultural contact.

What these anecdotal reports demonstrate is the degree to which, by the middle third of the nineteenth century, discussions of extinction had become part of a broader political and cultural discourse in Britain and elsewhere. However, while discussions of extinction expanded beyond the original geological and paleontological contexts explored in chapter 1 of this book, political and cultural understandings of extinction at this time were not easily separable from contemporary debates in elite scientific circles, but rather formed a broader imaginary that extended to political and popular discussions about race and empire. From the 1830s and 1840s onward, it became more and more common, for example, to find discussions of the "extinction" of "primitive" tribes encountered by Europeans in newspaper articles and parliamentary reports as the British and French expanded their imperial holdings, and in the United States as westward expansion intensified conflict between settlers and Native Americans. At the same time, questions about human race and social progress increasingly became implicated in scientific arguments about biological extinction. In scientific contexts, ideas about extinction played a central role in the emerging disciplines of anthropology, ecology, geography, and sociology, as well as in biology and paleontology. Lyell's influential interpretation of extinction as consistent with a balance of nature in dynamic equilibrium gave it a kind of positive moral valence: extinction was necessary, and even *good*, for the maintenance of a stable economy of nature. The flip side of this notion, of course, is that those species that became extinct were somehow at fault since they had failed to adapt to their changing environments, and that it was only just that they should make way for those that could survive. In a political context, the supposed inevitability of extinction reinforced cultural attitudes about social progress that justified the spread of European civilization, even at the cost of assimilation, subjugation, and even extermination of native peoples. Inevitably, these scientific and cultural attitudes about extinction contributed to how diversity—and especially the diversity of non-European peoples and cultures—was valued.

This view that extinction has an intrinsic, progressive valence, implicit in Lyell, was made much more explicit in Darwin's evolutionary theory, which provided a mechanism—natural selection—that explained and naturalized survival and failure. But Darwin also transformed Lyell by emphasizing the local instability of environments—how the constant struggle for existence made the toehold on survival of every individual and species tenuous—while at the same time maintaining that nature was an endlessly self-renewing source of new diversity. In the *Origin of Species*, Darwin treated the relationship between extinction and the emergence of new species as a kind of dynamic equilibrium, and argued that the total number of living species remained stable over time. However, whereas Lyell had adapted the principle of equilibrium to an essentially static chain of being, Darwin made it central to his theory of evolution, as a logical consequence of the central mechanism of natural selection. "Balance," for Darwin, meant that while the actors may be constantly entering and departing the stage, broadly speaking the play remains the same. Extinction was central to his particular concept of the economy of nature: If natural selection is the principle that favors those individuals—and ultimately species— best suited to survival and reproduction, then extinction is simply the fate of those who cannot successfully compete. Again and again in the *Origin* Darwin reinforced this point, explaining that "it inevitably follows, that as new species in the course of time are formed through natural selection, others will become rarer and rarer, and finally extinct," and that since a species is "maintained by having some advantage over those with which it comes into competition . . . the consequent extinction of less-favoured forms almost inevitably follows" (Darwin 1859, 110, 320). Because of the Malthusian principle of limited resources and fierce competition, natural selection is essentially a zero-sum game in which the number of winners will always be balanced by an equal number of losers. Far from viewing it as mysterious or anathema, Darwin conceived extinction as an essential process for keeping nature in a healthy balance.

It is impossible to avoid reading discussions of extinction in Darwin and other nineteenth-century authors in the broader context of Victorian beliefs about competition, social progress, racial hierarchy, and imperialism, and I will explore some of these connections in this chapter. However, I do not want to argue that Darwin or other Victorian scientists adopted their views about extinction *because* of prevailing social values, or vice versa. The true picture is much more complicated than that; cultural and biological understandings and valuations of extinction developed in tandem, each reinforcing the other in a complex chicken-and-egg relationship. This is true of the broader relationship between scientific and cultural values as well. During the latter decades of the nineteenth century, Darwin's biological theory was sometimes rather crudely applied to social problems, and this phenomenon has been labeled "social Darwinism." But current historical scholarship has called the stability and reliability of that label into question, and I do not find social Darwinism to be a very accurate or useful explanatory category.[4] Attempts to reduce either the social to the biological or the biological to the social are doomed to failure; Darwin himself drew heavily on social and economic theory when constructing his biological arguments, and it might be just as reasonable to call him a "biological Malthusian." But the point is really that the reductive approach is not profitable. In the Victorian era, as today, scientists were part of their culture, and culture was reflected in science.

Race and Extinction before Darwin

Darwin's view, as has been pointed out ever since 1859, appears to endorse a ruthlessly competitive view of nature, and his view of extinction seems to conceive of extinction as an inevitable and even progressive force. Darwin's and Lyell's views about extinction were part of a much larger nineteenth-century discourse related to British and European imperial expansion, and in particular to questions about the justification for exploiting and eradicating the native peoples, flora, and fauna encountered during colonization. British, European, and American expansion was often underwritten by an explicit belief that it was justifiable to subjugate and even exterminate so-called savage tribes because such "races" were doomed anyway by the inexorable logic of biology. This attitude also clearly implicates nineteenth-century European biological and anthropological theories of race, which experienced an explosion of interest during the same period that biological ideas about extinction were being developed. Darwin's ideas certainly contributed to this broader discourse or imaginary, and Darwin himself had much to say about racial hierarchies, social progress, and human extinction. But in many ways Victorian debates about race and extinction were independent of Darwin, and Darwin's views and influence were part of a larger context that predated the publication of *Origin of Species* in 1859. What I will emphasize later in this chapter, however, is the way in which Darwin transformed many of these older tropes in the context of his theory of evolution via natural selection.

The history of European biological ideas about race is long and complex, and this is not the place to try to enter into it deeply. In the eighteenth century, European theorists generally did not treat human biological difference as a matter of innate physiology or heredity, instead favoring "environmental" explanations for apparent differences between human groups. The human race was assumed to have descended from an original stock—often from the literal Adam and Eve—and existing "races" were groups of descendants that had been subject to greater or lesser "degeneration," depending on degrees of geographic isolation and cultural factors. However, at the beginning of the nineteenth century, new hereditarian or racialist ideas became popular in Europe, thanks in part to the emergence of comparative anatomy, which provided "evidence" of supposedly innate physical differences between human groups. The most common physical markers of race were cranial capacity and skull physiognomy, which were assumed to correlate directly to intelligence. European biologists and anthropologists tried to prove that this measure could be used to arrange the human races into a hierarchy, with Australian aboriginals, Africans, and other peoples native to territories under European imperial domination at the bottom, and Europeans themselves unsurprisingly at the very top. The most extreme versions of these arguments—exemplified, for example, in Josiah Knox and Francis Gliddon's *The Races of Man* (1850)—argued that individual human races were actually distinct species, and that the "lower" races were more closely related to apes than were the "higher" ones. These kinds of biological arguments could endorse all sorts of political ones, including the justification for owning slaves.[5]

There was considerable scientific controversy surrounding this issue, which is most often characterized as the debate between "polygenists," those who believed in multiple species of humanity, and "monogenists," those who held that humans were a single species, and that individual races were mere "varieties." Ultimately, Darwin—and Western scientific opinion—came down on the side of monogenesis, in part because it was argued that an evolutionary framework allowed insufficient time for human beings to have differentiated into separate species. Darwin and many contemporaries also found fault with much of the purported physical evidence for the polygenist position, and more broadly objected to characterizations of extreme innate mental and physical differences between races, as well as to the political agendas they often served, such as slavery. At the same time, however, even the more liberal members of the scientific elite believed that there was justification for ranking races or civilizations in some kind of hierarchy, even if it was based on cultural rather than innate differences.

One topic of frequent discussion both before and after 1859 was whether the "lower" races—again, judged either in hereditarian or cultural terms—were "doomed" to inevitable extinction by the spread of European imperialism. This question was often asked explicitly to justify European expansion or to assuage guilty consciences about its consequences, and it was a central theme in the Victorian extinction imaginary. In France, for example, members of the Paris Geographical and Ethnographical Societies provided racial justifications for colonial expansion as early as the 1820s, based on supposedly scientific study of cultural, though not always hereditarian, racial differences. In their milder forms, these arguments justified cultural assimilation of native peoples as a humanitarian "civilizing" process that would benefit both Europeans and natives. However, ethnographers such as René-Primevére Lesson and Jules-Sébastien-César Dumont d'Urville argued that some peoples—Australian and Polynesian aboriginals, for example—were "uncivilizable" and therefore inevitably doomed by contact with Europeans. While there might have been some passing regret, many geographers and ethnographers nonetheless had few scruples about urging expansion, arguing that extinction was the "natural" course of things.[6]

Many of the same arguments were put forward in Britain around the same time, as the excerpts from the parliamentary reports with which this chapter began show. Darwin himself first entertained these ideas while aboard HMS *Beagle* during the early 1830s (well before his evolutionary ideas were fully developed), remarking in his account of that voyage that "wherever the European has trod, death seems to pursue the aboriginal" in the form of disease, and observing that "the varieties of man seem to act on each other in the same way as different species of animals—the stronger always extirpating the weaker" (Darwin 1909, 459). Well before Darwin's evolutionary ideas were published, however, James Cowles Prichard had written an essay in the *Edinburgh New Philosophical Journal* titled "On the Extinction of the Human Races," where he argued that human extinction occurs naturally when tribes or races of people are placed in natural competition. Prichard, an Edinburgh physician, was a committed monogenist, and his stance toward native peoples was progressive and humanitarian for its time. For example, he lamented the "whole races [that] have become extinct during the few centuries which have elapsed since the modern system of colonialization have commenced," and urged his fellow scientists "to take up seriously the consideration, whether any thing can be done effectually to prevent the extermination of the aboriginal tribes" (Prichard 1840, 168, 170). Nonetheless, his prognosis was fatalistic:

> Wherever Europeans have settled, their arrival has been the harbinger of extermination to the native tribes. Whenever the simple pastoral tribes come into relations with the more civilized agricultural nations, the al-

lotted time of their destruction is at hand; and this seems to have been the case from the time when the first shepherd fell by the hand of the first tiller of the soil. . . . It may be calculated that these calamities . . . are to be accelerated in their progress; and it may happen that, in the course of another century, the aboriginal nations of most parts of the world will have ceased entirely to exist (Prichard 1840, 169).

The most Prichard could offer as a solution was to urge that "if Christian nations think it not their duty to interpose and save the numerous tribes of their own species from utter extermination, it is of the gravest importance, in a philosophical point of view, to obtain much more extensive information than we now possess of their physical and moral characters" (Prichard 1840, 169–70).

This attitude of regretful fatalism is little different from Archdeacon Broughton's comments to Parliament about the aborigines of New South Wales; and it is found in other published works of the same time, such as Charles Hamilton Smith's *Natural History of the Human Species* (1851). Smith, a monogenist like Prichard, discussed the inevitability of human extinction through competition as regrettable but inevitable, stating that while "it would be revolting to believe that the less gifted tribes were predestined to perish beneath the conquering and all-absorbing covetousness of European civilization, without an enormous load of responsibility resting on the perpetrators," nonetheless "their fate appears to be sealed in many quarters, and seems, by a preordained law, to be an effect of more mysterious import than human reason can grasp" (Smith 1851, 207). Smith's essay also made use of the same analogy between individuals and groups found in Brocchi and Lyell, arguing that, "as it is with individual life, so families, tribes, and nations, most likely even races, pass away." He argued that this process was inevitable and even natural, since "their tenure is only provisional, until the typical form appears, when they are extinguished, or found to abandon all open territories not positively assigned them by nature, to make room for those to whom they are genial" (Smith 1851, 175). In this way, whatever humanitarian regret Hamilton expressed was balanced by the fatalistic perception that extinction of "inferior" races was both inevitable and "lawful." In fact, for Smith the extinction of a particular race was tantamount to proof of its inferiority; as he described the plight of Native Americans, "The decay, amounting to prospective extinction" was in fact "a further proof that they are not a typical [that is, well-adapted] people," since nontypical people were "alone liable to annihilation, or to entire absorption" (Smith 1851, 276).

Smith's neat tautology well represents typical European scientific sentiment at the middle of the nineteenth century: whether regrettable or not, the inevitable fate of non-European peoples in the face of European contact was complete cultural assimilation at best, and utter extinction at worst. Furthermore, this process was often characterized as a "law of nature," assuaging potentially uneasy consciences and implicitly endorsing the politics of imperialism. While it may well have been the case that individual authors would have supported imperial expansion in any event, I want to emphasize that these biological justifications were not ad hoc. They were based not only on the racialized anthropology and ethnography of the day, but also on the leading theories of biological extinction—such as Lyell's—on which they explicitly drew. Indeed, Lyell himself and—as we will see shortly—Darwin, contributed directly to this discourse of racial extinction, which flowed naturally from their larger theoretical frameworks.

A final important point is what these approaches to the problem of extinction say about the value placed on diversity. Prichard's lament and Hamilton's qualms seem to have little to do with regret over the diminishment of absolute human cultural diversity or variety. After all, either man would have been quite satisfied with a "civilizing process" that involved the complete cultural assimilation of native peoples. One might call the outcome of this anticipated civilizing process a kind of "soft extinction," where the natives themselves may be physically spared from extermination but their culture would vanish with little regret. Rather, the dominant sentiment appears to have been pity, and at most a rather selfish regard for the loss of cultural data that could help Europeans construct comprehensive anthropological or ethnographic theories. Cultural diversity as such was simply not valued.

Darwin on Competition, Extinction, and the Economy of Nature

While Darwin certainly did not begin the discourse about extinction, competition, and the balance of nature that I have been discussing, thanks to the significance and notoriety of his evolutionary works his theory of descent with modification by natural selection became the lens through which much of the subsequent debate was filtered. Darwin was well aware of earlier literature on both extinction and the economy of nature, which had a significant influence on his own thought. As many historians have demonstrated, the period between 1830 and 1842 was a formative time for him; this includes his voyage around the world on HMS *Beagle* from 1830 to 1835, the composition of his early notebooks recording his developing ideas about natural selection and transmutation shortly after his return, and the drafting of an initial sketch of his theory. While Darwin's ideas were clearly influenced by direct experience of the geology, flora, and fauna of South America and Oceania during his travels, much of his time on the voyage was spent simply reading; and from the lists and private notes he made, it is possible to have a fairly clear idea of which books he read, and how they helped shape his theory.

It is fairly well documented that when Darwin first set out on the *Beagle* he was not committed to any theory of transmutation. Experiences such as his encounter with the fossil remains of large vertebrates in South America and with the finches and tortoises of the Galapagos Islands had a profound influence on his thinking about the variation, change, and historicity of organisms. But these direct experiences were also shaped by his reading—in particular, his reading of the first two volumes of Lyell's *Principles of Geology*, which helped provide a framework for what he was witnessing. From Lyell in particular he became convinced that the earth was very old, and that its geology was shaped by gradual, dynamic processes that over time could raise or lower continents and build mountain ranges. To Darwin, this evidence of slow geological change seemed at odds with the catastrophic revolutions proposed by Cuvier, whom Darwin also read during his trip. The private notebooks Darwin kept during the voyage demonstrate that he was quite receptive to Lyell's views about extinction, as expressed in volume 2 of *Principles of Geology*. For example, in a short sketch written in 1835, Darwin reflected on the succession of fossil vertebrates he had observed in South America:

> With respect then to the death of species of Terrestrial Mammalia in the S. part of S. America. I am strongly inclined to reject the action of any sudden debacle. — Indeed the very numbers of the remains render it to me more probable that they are owing to a succession of deaths, after the ordinary course of nature. — As Mr Lyell (a) supposes Species may perish as well as individuals; to the arguments he adduces. I hope the Cavia of B. Blanca will be one more small instance, of at least a relation of certain genera with certain districts of the earth. This co-relation to my mind renders the gradual birth & death of species more probable (Darwin 1835).[7]

This statement is a reference to Brocchi's analogy between individual and species life spans, which Darwin had learned about through Lyell. There is some scholarly disagreement about the exact significance of this passage, but it is clear that on some level Darwin was endorsing the analogy itself, if not Brocchi's internalist explanation for it. Nevertheless, Darwin went on to write that "If ~~gradual deaths~~ the existence of species is allowed, each according to its kind, we must suppose deaths to follow ~~one after~~ at different epochs, & then successive births must re-people the globe or the number of its inhabitants has varied exceeding at different periods. — A ~~fact~~ supposition in contradiction to the fitness ~~wit~~ which the Author of Nature has now established." At this early stage in his thinking, Darwin agreed with Lyell that the total diversity of life at any given time should not change — this natural balance of nature is established by "the Author of Nature." In order to maintain an equilibrium, species births must more-or-less exactly match their deaths, and evidence for this is found in the succession of different forms of large mammals in the geology of South America. Darwin also accepted that this process of extinction and replacement must have a natural cause — as opposed, say, to divine intervention and special creation. He was inclined to accept Lyell's environmental explanation for species extinction, but this did not explain how the new forms appeared, or why those forms often seemed to be closely related to their extinct predecessors. As some observers have argued, this was a watershed moment for Darwin: he was now on the path to realizing that transmutation would effectively answer the problem.[8]

These passages also show that, while composing his initial notes aboard the *Beagle*, Darwin was committed to the idea that geologic and organic change was marked by a natural balance or equilibrium. This balance was manifested in two ways: in the adaptedness of organisms to their environments, where every available "station" would be filled by a creature adapted to survive in it, and in the gradual replacement of species as those environments slowly changed. When Darwin returned to England in 1835, he began working through the evidence he had gathered during his voyage, and extinction became central to these reflections. In an entry marked January 1834, he had remarked in his *Beagle* notebook that "we are so profoundly ignorant concerning the physiological relations, on which the life, and even health . . . of any existing species depends, that we argue with still less safety about either the life or death of any extinct kind." While he speculated that "simple relations" such as change in climate or predation may explain "the succession of races," he nonetheless concluded, "All that at present can be said with certainty, is that, as with the individual, so with the species, the hour of life has run its course, and is spent" (Darwin 1839, 211–12). Here Darwin recognized extinctions as a normal feature of the economy of nature, but was ambivalent about their causes or broader significance.

In the notebooks he kept between 1836 and 1838, Darwin began to home in on a more concrete explanation for this process of dynamic replacement. In his early "Red Notebook" of 1836–37 he remarked, "There is no more wonder in extinction of species than of individual." This was a variation on a statement he would make in many of his later works: Extinction is a common and natural occurrence.[9] A year later, he described the "quantity of life" on earth as a fluctuating balance, depending upon the relationship between organisms and the dominant environments they occupied, but remarked that "this perhaps on long average equal" (Darwin, Notebook C 1838, 147e). An important moment came in September of 1838, when Darwin read the sixth edition of Thomas Malthus's *Essay on the Principle of Population.* As Darwin would recall in his *Autobiography*, Malthus provided the inspiration for natural selection: "Being well prepared to appreciate the struggle for existence which everywhere goes on from long-continued observation of the habits of animals and plants, it at once struck me that under these circumstances favourable variations would tend to be preserved, and unfavourable ones to be destroyed. The result of this would be the formation of new species. Here, then, I had at last got a theory by which to work" (Darwin 1887, 83). It may well be the case that Darwin drew inspiration for natural selection from Malthus, but historians have also pointed out that he had already encountered sources—Lyell and De Candole, in particular—on the importance of competition or "struggle" in nature, and some have questioned the continuity between Malthus's and Darwin's views on the subject.[10]

What is clear, though, is that Darwin's reading of Malthus had a clarifying effect on his understanding of the role of competition in maintaining the stability of what we would now call ecological relationships. As Donald Worster argues in his comprehensive account of the history of ecology, *Nature's Economy*, it was "the single most important event in the history of Anglo-American ecological thought" (Worster 1994, 149). Unlike many of his late-eighteenth-century contemporaries, Malthus believed that an equilibrium was maintained in nature through fierce competition for scarce resources, rather than as the product of a harmonious, beneficent plan. For Malthus, the economy of nature was preserved by an imbalance between population and resources: "Necessity, that imperious, all-pervading law of nature" ensured that only as many individuals as could be supported were able to survive.[11] The influence on Darwin's thinking about the role of competition in the economy of nature was immediate: in a famous passage in his "Notebook D" of 1838, he wrote:

> One may say there is a force like a hundred thousand wedges trying to force <into> every kind of adapted structure into the gaps <of> in the oeconomy of Nature, or rather forming gaps by thrusting out weaker ones. <<The final cause of all this wedgings, must be to sort out proper structure & adapt it to change (Darwin Notebook D 1838, 135e).

This passage is striking because it introduces a "force" that drives extinction, something much more active and dramatic than what Lyell had described as the gradual pressure of slow environmental change. The force Darwin mentioned was simply competition but, filtered through his reading of Malthus, he now saw it as something that actively maintained the economy of nature. But Darwin was also developing a conceptualization of "economy" that was less static than Malthus's or Lyell's. Whereas Malthus had understood relationships to be essentially fixed, and Lyell saw change as a cyclical process that slowly fluctuated between fixed boundaries, Darwin was opening the door to a notion of "balance" that admitted dramatic and permanent change (via transmutation) and also acknowledged the instability and tenuousness of local environments, while at a deeper level affirming a commitment to a view of nature as self-generating and self-renewing.

This stage also marked a much more confident treatment of extinction in Darwin's writings. For example, in the first edition of his *Journal of Researches*, his record of his experiences aboard HMS *Beagle*, published in 1839 but composed earlier, Darwin recounted his observation of evidence of extinct South American vertebrates in the ambivalent terms quoted above (the "January 1834" entry). However, in the second edition of *Journal of Researches*, published in 1845, his tone had changed markedly. In the same section ("January 1834"), he now confidently asserted, "Certainly, no fact in the long history of the world is so startling as the wide and repeated extermination of its inhabitants. Nevertheless, if we consider the subject under another point of view, it will appear less perplexing" (Darwin 1845, 174–75). The point of view Darwin was referring to was clearly the Malthusian dynamic of a "geometrical" rate of population increase, combined with a constant availability of resources. All the earlier passages expressing ambivalence towards extinction's causes had now been cut, and Darwin adopted the "Lyellian" perspective that "an action going on, on every side of us, and yet barely appreciable, might surely be carried further, without exciting our observation" (Darwin 1845, 176). Noting Lyell's dictum that "rarity is the precursor to extinction," he argued that extinction of a species ought to excite no more wonder than the sickness and death of an individual organism.

This last argument was something Darwin had worked though in two important handwritten "sketches" of his developing theory, in 1842 and 1844 respectively. In the 1842 sketch he explicitly adopted the Brocchian/Lyellian analogy between the individual and the species, writing, "It accords with what we know of the law impressed on matter by the Creator, that the creation and extinction of forms, like the birth and death of individuals should be the effect of secondary <laws> means" (Darwin 1842).[12] In the 1845 sketch, he confronted the problem of apparent mass extinctions, which he argued were artifacts of gaps in the fossil record. While he acknowledged cases in which "extinction might be locally sudden"—for example, after a local flood—he maintained that there were "no grounds whatever" to support Cuverian cycles of catastrophes: "All [evidence] seem[s] to show that the extinction of the several classes and renewal of species does not depend on general catastrophes, but on the particular relations of the several classes to the conditions to which they are exposed" (Darwin 1844).[13] The economy of nature was such that extinctions would be generally compensated for by the creation of an equal number of new species. In a number of places in this later sketch, Darwin returned to the analogy between the "births" and "deaths" of individuals and species. Importantly, this "balance" was conceived as the system's overall tendency to maintain consistent levels of diversity while experiencing constant extinction and replacement.

This is generally the position that Darwin took in the first edition of *Origin of Species*, where extinction through competitive replacement became enshrined in his principle of natural selection. As he put it, "It follows [from natural selection] that as each selected and favoured form increases in number, so will the less favoured forms decrease and become rare. . . . It inevitably follows, that as new species in the course of time are formed through natural selection, others will become rarer and rarer, and finally extinct" (Darwin 1859, 109–10). Darwin also made it clear that this was a matter of natural law, following the inexorable logic of natural selection: a species is "maintained by having some advantage over those with which it comes into competition; and the consequent extinction of less-favoured forms almost inevitably follows" (Darwin 1859, 320). This would, of course, be a slow and gradual process. In the *Origin*, as in his 1844 sketch, Darwin had no place for "the old notion of all the inhabitants of the earth having been swept away at successive periods by catastrophes"; rather, "species and groups of species gradually disappear, one after the other, first from one spot, then from another, and finally from the world" (Darwin 1858, 317–18).

Darwin's views about extinction in the *Origin* also contributed to his position on the economy of nature, and on the value of a division of labor among organisms that he termed "divergence." He observed:

> Battle within battle must ever be recurring with varying success; and yet in the long-run the forces are so nicely balanced, that the face of nature remains uniform for long periods of time, though assuredly the merest trifle would often give victory to one organic being over another. Nevertheless so profound is our ignorance, and so high our presumption, that we marvel when we hear of the extinction of an organic being; and as we do not see the cause, we invoke cataclysms to desolate the world, or invent laws on the duration of the forms of life! (Darwin 1859, 73).

However, despite being in constant motion, the net diversity of life is not significantly affected by the extinction of old and the creation of new species. "Everyone has heard that when an American forest is cut down," he remarked elsewhere, "a very different vegetation springs up; but it has been observed that the trees now growing on the ancient Indian mounds, in the Southern United States, display the same beautiful diversity and proportion of kinds as in the surrounding virgin forests." In other words, nature's inherent fecundity ensures that there will always be new forms standing by to replace the old ones, and that those new species will survive if they maintain a competitive advantage with their environments. Diversity (or "divergence") allows organisms to "be better enabled to seize on many and widely diversified places in the polity of nature, and so be enabled to increase in numbers" (Darwin 1859, 112).

Darwin expanded on these ideas in the sixth edition of *Origin*, published in 1872, where he explicitly argued that extinction and diversification remain in harmonious, though dynamic, balance. While he acknowledged that "there seems at first sight no limit to the amount of profitable diversification of structure, and therefore no limit to the number of species which might be produced," nonetheless "geology shows us, that from the early part of the tertiary period the number of species of shells, and that from the middle part of this same period the number of mammals, has not greatly or at all increased" (Darwin 1872, 101–2). He concluded ultimately that "thus the appearance of new forms and the disappearance of old forms, both those naturally and those artificially produced, are bound together. . . . We know that species have not gone on indefinitely increasing, at least during the later geological epochs, so that, looking to later times, we may believe that the production of new forms has caused the extinction of about the same number of old forms" (Darwin 1872, 296).

Remarkably absent in any edition of the *Origin* is the sense that Darwin viewed biological diversity in the way that scientists do today. Diversity—or "biodiversity," in the term currently used—is an enormously complicated and often slippery concept, as a number of authors have pointed out.[14] Traditionally, biological diversity is understood as a measure of "species richness"; that is, the absolute number of different kinds of organisms in a particular environment. But that definition hardly does justice to the complex and nuanced way biodiversity is understood in discussions of ecology, conservation biology, paleontology, and other related contexts, to say nothing of political, economic, and cultural discourse. Nor does it take account of the many potentially problematic assumptions inherent in such a limited definition— from very basic taxonomic questions about how biological entities are defined, to complex and culturally laden associations of diversity with utilitarian, theological, and philosophical schemes of valuation. According to one recent definition, biological diversity is "the variety of living organisms; the biological complexes in which they occur, and the ways in which they interact with each other and the physical environment" (Groves et al. 2002, 500). Without wading too deeply into this debate, these problems will be discussed more directly in chapter 6 of this book.

What is safe to say is that Darwin did not use the term in a way that reflected even the basic definition of diversity as species richness. The word does appear as a noun (as opposed to adjectival forms like "diverse" or "diversified") some eighteen times in the text of the first edition of the *Origin*, but in every case it is used merely as a synonym for "variety." For example, the word appears several times in the first chapter, "Variation under Domestication," where Darwin discusses the "diversity" of different breeds of pigeons, or flowers in a garden, or fruit in an orchard.[15] Darwin also sometimes used the term "diversity" when discussing differences in the morphology or structure of limbs and organs, as in "diversity in the shape of the pelvis of birds" (127), or "graduated diversity in the eyes of living crustaceans" (188), or "infinite diversity in structure and function of the mouths of insects" (436). For Darwin, diversity was essentially a comparative term, meant to indicate an amount or degree of difference between the features of organisms. It does not convey a sense of ecological interdependence, nor is it usually presented as a broader phenomenon that is threatened or in need of preservation. In fact, in the two instances in the *Origin* where Darwin invoked "diversity" in a somewhat broader sense, it was presented as an example of how "beautiful" or "harmonious" the balance of nature is.[16]

While "diversity" was not central in Darwin's theory, the term "divergence" was, as in "the principle of divergence," which is the principle by which natural selection favors a multiplicity of different adaptations that allow organisms "to be better enabled to seize on many and widely diversified places in the polity of nature, and so be enabled to increase in numbers" (Darwin 1859, 112). This is what Darwin considered akin to a division of labor: "The advantage of diversification of the inhabitants of the same region is, in fact, the same as that of the physiological division of labour in the organs of the same individual body" (Darwin 1859, 115).[17] But even this concept was problematic for Darwin. As Worster argues, "Darwin never seemed able to focus on these implications of the principle of divergence . . . for they complicated and even contradicted the emphasis he placed on competitive replacement" (Worster 1994, 162).

Indeed, Darwin barely seemed to consider the possibility that nature could ever run out of material—diversity—with which to populate its many "stations." To the extent that he recognized something like biological diversity in nature, he regarded it as an endlessly renewing resource. This attitude reflects the older notion of "plentitude" in nature, associated with Linnaeus and other theologically inspired naturalists, as well as Lyell's interpretation of geological history as a dynamic equilibrium. What Darwin added was the regular cycle of extinction and speciation, which made Darwin's view of nature considerably more transient than earlier conceptions of a static balance or economy. But beneath this constant change is a fundamental underlying stability, provided thanks to nature's capacity for endless self-generation of more diversity. The issue, then, isn't whether Darwin recognized or thought natural variety was important—he certainly did—but whether he thought diversity itself could be diminished by extinction, and nature's stability could thus be threatened, which he did not. Competition and replacement were, for Darwin, the engine that drove the progressive improvement of the natural system and maintained the economy of nature. Far from seeing diversity as something to be conserved, he viewed it as essentially the fuel for that engine, the source of continued competition, selection, and extinction. The extinction of a species somewhere *always* opens up the possibility of a new one somewhere else; this was as much a "law" of nature as anything to be found in the *Origin*. The idea that nature exists in a harmonious, unchanging balance may have been upset, at the end of the eighteenth century, by authors such as Malthus and Cuvier, who suggested that competition and the specter of extinction were an inherent part of the natural order. But Darwin's message was, essentially, that struggle and even extinction were positive forces, in the long view—thus soothing the anxieties of Victorians about their own impact on the world. The world may be subject to constant change, but faith in the ultimate constancy of nature was not shaken.

Extinction and Cultural Progress

Famously, Darwin said virtually nothing in the *Origin* about the implications of natural selection and evolution for humans, beyond the cryptic statement that "In the distant future. . . . Light will be thrown on the origin of man and his history" (Darwin 1859, 488). As we have already seen, however, other naturalists including Lyell had already extended the study of extinction as a natural process to considerations of its significance for human culture and civilizations. The theme of "inevitable" racial extinction was already a well established trope by the 1850s, and there is plenty of evidence that Darwin was well aware of the potential implications for his own developing theory. In observations of his experiences in New South Wales in January of 1836, Darwin remarked in his *Journal of Researches*:

> Wherever the European has trod, death seems to pursue the aboriginal. We may look to the wide extent of the Americas, Polynesia, the Cape of Good Hope, and Australia, and we shall find the same result. Nor is it the white man alone, that thus acts the destroyer; the Polynesian of Malay extraction has in parts of the East Indian archipelago, thus driven before him the dark-coloured native. The varieties of man seem to act on each other; in the same way as different species of animals—the stronger always extirpating the weaker (Darwin 1839, 520).

This shows that, as Darwin was developing his evolutionary ideas, he was conscious of the analogies that could be drawn between extinction in the human and nonhuman spheres. In a letter to Lyell in October 1859 he observed that naturalized European plants in South America "conquer the aborigines," and he discussed a similar phenomenon whereby the "most intellectual individuals of a species" might be favorably selected: "I look at this process as now going on with the races of man; the less intellectual races being exterminated" (Darwin 1839, 520).

When it came time to publish his extension of his theory of evolution to human beings in *Descent of Man* (1871), Darwin took much the same line. He accepted the extinction of human races as "historically known events," and bluntly stated that "extinction follows chiefly from the competition of tribe with tribe, and race with race. . . . When civilized nations come into contact with barbarians the struggle is short, except when a deadly climate gives its aid to the native race" (Darwin 1871, 236–38). While Darwin was certainly not shy about using terms such as "barbarians" to describe indigenous peoples, he drew back from explicitly endorsing or excusing violent extermination of native peoples as a consequence of European imperial expansion. Nonetheless, his more general statements on the subject made it clear that he felt natural competition between races could have a generally "improving" effect on the human species. He argued, for example, that while a "tribe" of "selfish and contentious people" might have a temporary advantage over more peaceful ones, ultimately groups of people with higher "social and moral qualities would tend slowly to advance and be diffused throughout the world" (Darwin 1871, 162–63). Darwin believed, perhaps rather optimistically, that this process would take place chiefly without bloodshed, noting that "at the present day civilized nations are everywhere supplanting barbarous nations . . . and they succeed mainly, though not exclusively, through their arts, which are the products of the intellect" (Darwin 1871, 160).

In this regard, Darwin's published views were anticipated by a decade or more by authors who are generally considered part of the "Darwinist" camp—most prominently Alfred Russell Wallace and Herbert Spencer. In 1864, Wallace published an essay titled "The Origin of Human Races and the Antiquity of Man Deduced from the Theory of Natural Selection," which was based on an address he had given to the Anthropological Society of London. The primary purpose of Wallace's lecture was to argue for the applicability of natural selection to human evolution, but in the process he gave considerable attention to extinction. Like Darwin, Wallace was optimistic that "tribes in which such [refined] mental and moral qualities were predominant, would therefore have an advantage in the struggle for existence over other tribes," from which it would inevitably follow that "the better and higher specimens of our race would therefore increase and spread, the lower and more brutal would give way and successively die out" (Wallace 1864, clxii–clxiv). This generally gives the impression that with humans, unlike "lower" animals, competition would tend to be intellectual rather than violent, and that "the power of natural selection . . . must ever lead to the more perfect adaptation of man's higher faculties to the conditions of surrounding nature, and to the exigencies of the social state" (Wallace 1864, clxix).

At the same time, Wallace made no bones about whom he considered to be the "superior" and "inferior" races, nor about what the inevitable result of European expansion would be:

It is the same great law of "*the preservation of favored races in the struggle for life*," which leads to the inevitable extinction of all those low and mentally undeveloped populations with which Europeans come in contact. The red Indian in North America, and in Brazil; the Tasmanian, Australian and New Zealander in the southern hemisphere, die out, not from any special cause, but from the inevitable effects of an unequal mental and physical struggle. The mental and moral, as well as the physical qualities of the European are superior . . . [and] enable him when in contact with the savage man, to conquer in the struggle for existence, and to increase at his expense, just as the more favorable increase at the expense of the less favorable varieties in the animal and vegetable kingdoms, just as the weeds of Europe overrun North America and Australia, extinguishing native production by the inherent vigour of their organisation, and by their greater capacity for existence and multiplication" (Wallace 1864, clxiv–clxv).

What is especially striking about this passage—aside from its rather blithe attitude towards the extinction of human beings—is the implied message about diversity. In Wallace's view of social evolution, the net result of "more perfect adaptation" is *less* rather than more diversity. Wallace went on to argue that mental abilities would continue to evolve "till the world is again inhabited by a single homogeneous race, no individual of which will be inferior to the noblest specimens of existing humanity." The end result, in Wallace's view, would be a utopian society with perfect freedom, universal altruism, and no need for laws, governments, or police; in short, as he put it, the earth will have been converted "into as bright a paradise as ever haunted the dreams of seer or poet" (Wallace 1864, clxx). Darwin sent Wallace an enthusiastic letter in response, which complimented the assertion of mental evolution as the "great leading idea" of the essay, and commented, "The latter part of the paper [e.g., the section on human competition, extinction, and progress] I can designate only as grand & most eloquently done" (Darwin to Wallace, 28 May 1864). In his reply, Wallace thanked Darwin but also—famously—asserted, "As to the theory of '*Natural Selection*' itself, I shall always maintain it to be actually yours & your's [*sic*] only" (Wallace to Darwin, 29 May 1864).

However, another source of inspiration for Wallace, which he acknowledged with a footnote in his 1864 essay, was Herbert Spencer's *Social Statics* (1851). Indeed, Spencer had been writing about social evolution for several years before the publication of Darwin's *Origin*, and his views about the role of competition among humans as a mechanism for social progress anticipated Darwin's own. It is worth noting, however, that Spencer's vision of evolution was markedly more progressionist than Darwin's; Spencer unabashedly believed that in both biology and society, evolution produced "better" outcomes. In *Social Statics* and elsewhere, Spencer consistently maintained that competition among human groups or races was not only natural but good, even though the consequences for the less successful were invariably dire. "Inconvenience, suffering, and death, are the penalties attached to nature by ignorance," he wrote in *Social Statics*, "as well as to incompetence—and also the means of remedying these. . . . Partly by weeding out those of lowest development, and partly by subjecting those who remain to the never-ceasing discipline of experience, nature secures the growth of a race who shall both understand the conditions of excellence, and be able to act up to them" (Spencer 1851, 380). While Spencer, like Wallace, envisioned the resulting society as one in which less suffering and inequality would exist, he nonetheless acknowledged—a little too comfortably for even some of his contemporaries—that a certain amount of unpleasantness would precede that state.

In an 1852 essay in the *Westminster Review* titled "A Theory of Population, Deduced from the General Law of Animal Fertility," Spencer argued that "families and races" which tended to produce an "excess of fertility" without an accompanying "greater mental activity"—he singled out the Irish as a case in point—"are on the high road to extinction; and must necessarily be supplanted by those whom the pressure does so stimulate" (Spencer 1852, 35–36). And in his 1873 *The Study of Sociology*, he even more explicitly discussed the salutary effect of competition between races—even to the point of violent warfare—in raising the level of civilization:

> Warfare among men, like warfare among animals, has had a large share in raising their organizations to a higher stage. . . . In the first place, it has had

the effect of continually extirpating races which, for some reason or other, were least fitted to cope with the conditions of existence they were subject to. The killing-off of relatively-feeble tribes, or tribes relatively wanting in endurance, or courage, or sagacity, or power of co-operation, must have tended ever to maintain, and occasionally increase, the amounts of life-preserving powers possessed by men (Spencer 1873, 168).

However, Spencer also noted that in "higher societies," competition would tend toward cultural and economic battle rather than overt violence. This was, he argued, for the benefit of the elevated races, since "destructive activities" had "injurious effects on the moral natures of their members . . . which outweigh the benefits resulting from the extirpation of inferior races. . . . After this stage has been reached, the purifying process, continuing still an important one, remains to be carried on by an industrial war—by a competition of societies during which the best, emotionally, physically, and intellectually, spread most, and leave the lest capable to disappear gradually, from failing to leave a sufficiently-numerous posterity" (Spencer 1873, 173–74). The message, in other words, was that while advanced civilizations should refrain from actively exterminating native peoples, nature would eventually do the job for them.

The attitudes towards human extinction expressed by Wallace and Spencer (and implicitly by Lyell and Darwin) can be seen in varying degrees across a wide spectrum of literature in Britain, France, and the United States between 1860 and 1900. The belief that native peoples— "savages"—encountered by Europeans during colonial expansion were "doomed" to extinction has been documented by a number of historians as comprising an "extinction discourse" in later Victorian culture. This discourse, according to literary scholar Patrick Brantlinger, acted as "a powerful axis of ideas that has been hegemonic for countless European explorers, colonists, writers, artists, officials, missionaries, humanitarians, and anthropologists" (Brantlinger 2003, 190). The belief in the inevitable extinction of inferior races in many ways assuaged the guilt of European imperialists and, from one perspective, can be seen as a subset of contemporary race theory. At the same time, there was a great deal of heterogeneity and debate from the 1860s onward in anthropological and biological circles concerning the innate hereditary superiority of certain races over others. Racism, in other words, could take many forms, ranging from crude hereditarian theories of innate physical difference to more relativistic arguments about cultural superiority.[18] While theories of European racial superiority certainly contributed to this extinction discourse, I have tried to locate the roots of these attitudes more deeply in the biological understanding of extinction and its consequences developed by naturalists throughout the nineteenth century, because I think this gives us a broader and more interesting context in which to understand how broadly cultural and elite scientific values interpenetrated one another at this time.

Darwin's views about the essential unity of the human species—his rejection of polygenism—were widely shared by his contemporaries in the 1860s and afterwards. This did not mean, of course, that all races of people were considered equal. An earlier discourse of crude physiological hierarchy, based on "scientific" evidence from fields such as craniometry, came to be replaced by anthropological views that emphasized evolved cultural superiority, and which opened the possibility of the "elevation" of inferior races. Even so, this attitude was ultimately consistent with the dominant biological interpretation of extinction, since it still envisioned a competitive struggle between cultures in which the more "advanced" would inevitably triumph, even if it led to the peaceful assimilation rather than violent extermination of the vanquished. The "progress" that was envisioned by Darwin, Wallace, Spencer, and others might result in a kind of "soft" extinction, but either way the outcome would be the narrowing of human cultural diversity towards a "monoculture" based on the European ideal.

While it would be tedious to catalog statements by Europeans reflecting such views, a representative sampling will serve to emphasize the contribution of scientific extinction discourse into a broader cultural and political extinction imaginary. The extermination of the primitive was a trope that had broad cultural resonance in the later nineteenth century, even as a kind of poetic metaphor. Henry David Thoreau's *Walden* (1854), for example, speaks of seeking individual purity by overthrowing "savage" instincts and appetites: "We are conscious of an animal within us, which awakens in proportion as our higher nature slumbers. . . . Nature is hard to be overcome, but she must be overcome" (Thoreau 1854, 235–38). Overcoming nature could take a variety of forms: it could include extracting maximum economic yield from crops and livestock, establishing plantations of imported varieties in colonial territories, or improving the (European) standard of living through technological innovations in transport, communication, and medicine. And, quite explicitly, it could involve displacing, assimilating, or even eradicating populations of human beings that stood in the way of European expansion. The point, however, is that the received view of biological extinction contributed to this discourse by naturalizing Europeans' political and economic interests. It was not merely an ad hoc justification for rapacious imperialism—it was central to how European elites understood their role in the economy of nature.

Many authors quite explicitly associated Darwin's theory of natural selection—and by extension, his interpretation of extinction—with the kind of inevitable racial extinction that implicitly or explicitly justified European expansion. In an infamous 1868 *Fraser's Magazine* article titled "On the Failure of 'Natural Selection' in the Case of Man," the essayist and free-trade promoter William R. Greg noted that "in every part of the world, and in every instance, the result has been the same; the process of extinction is either completed or actively at work" (Greg 1868, 357). Greg's piece, which is often cited as an early inspiration for the eugenics movement, was broadly concerned with arguments about the role of social welfare laws in preserving "less fit" members of society. But he had no doubt that "the principle [natural selection] does not appear to fail in the case of *races* of men," where "the abler, the stronger, the more advanced, the finer in short, are still the favored ones," who "exterminate, govern, supersede, fight, eat, or work the inferior tribes out of existence." Greg explicitly cited Darwin as an authority for his views, and in particular glowingly endorsed Wallace's essay, which he quoted for more than a full page. Furthermore, Greg was clear that no number of moral qualms would make any difference for the outcome: "The process is quite as certain, and nearly as rapid, whether we are just or unjust; whether we use carefulness or cruelty. Everywhere the savage tribes of mankind die out at the contact of the civilized ones" (Greg 1868, 356).

Similar sentiments were common in European and American political and scientific publications and discussions throughout the remainder of the century. For example, the German physiologist and philosopher Friedrich Karl Christian Ludwig Büchner, a major exponent of anti-Romantic "scientific materialism," published a book that was translated into English in 1872 as *Man in the Past, Present, and Future*. Büchner was a fan of both Darwin's and Wallace's writings on the operation of natural selection in human societies, and his own book reflected many of the sentiments we have already observed—for instance, that "all backward branches of the great human family will by degrees disappear with but few exceptions under the pressure of civilized man." What is particularly interesting about Büchner's arguments is how he imagined that this process would homogenize the human species. Here he separated a "*reducing* movement" (i.e., extinction) from a "*differentiating* one," arguing that gradual extinction of inferior races would "superinduce a greater uniformity or similarity of mankind in all parts of the earth"; and he looked forward to "the time when a certain uniformity of culture and material conditions . . . will be diffused over the greater part of the inhabited and habitable part of our planet" (Büchner 1872, 153–54). Like Wallace, Büchner saw the reduction of diversity as a positive outcome, at least in regard to human culture.

Büchner's case reveals that "Darwinian" justifications of the inevitable extinction of humans through contact with Europeans were not limited to an Anglo-American context. Oscar Schmidt, a zoology professor at the University of Strasbourg, was an early German supporter of Darwin and the author of a popular book translated into English as *Darwin and the Doctrine of Descent and Darwinism* (1875). Schmidt's position on the human race veered towards polygenism; he wrote that "inferior human races exist—we may call them human species—which are related to the others, as are lower animals to higher." He also endorsed a fairly stark view of the consequences of European imperialism:

We are not to be misled by the contrary statements of missionaries and other philanthropists. . . . if we contemplate the ethnology and anthropology of savages, not from the standpoint of philanthropists and missionaries, but as cool and sober naturalists, destruction in the struggle for existence as a consequence of their retardation (itself regulated by the universal conditions of development), is the natural course of things (Schmidt 1875, 297–98).

In fact, it is quite difficult to find any European naturalists, anthropologists, or explorers in the later nineteenth century who did *not* regard racial extinction as the inevitable—and basically unpreventable—outcome of European expansion. We see this in Alfred Newton's exasperated comment that "It is seldom that any one but a Fennimore Cooper or a Charles Kingsley feels the romance that clings around the history of an expiring race. Most men—men of science especially—nowadays believe in the survival of the fittest, and are content to let the dead bury their dead" (Newton 1885, 546). Newton, who was the first professor of anatomy and zoology at Cambridge, was a passionate early activist for biological conservation. But even he separated extinction into two categories—"natural" versus "artificial" (i.e., caused by human agency)—and his concern was primarily directed toward a few specific examples of species, such as the great auk, that were being threatened by human activity. And even Newton's arguments for conservation were based mostly on the potential loss of valuable scientific information, rather than on any consideration of the value of biological diversity as such.[19]

Indeed, by the later nineteenth century there was increasing public interest in the role of human activities, especially hunting, in causing species extinctions. From the mid-1880s onward, letters and editorials in British and American newspapers reflected a growing popular concern with preserving individual species—or at least with recognizing the harmful effects of human agency after the fact. A number of letters were published in the *Times* of London from 1884 through the end of the century describing, in elegiac terms, the extinction or prospective extinction of North American elk and antelope, African elephants and aardvarks, and even larks and robins in Italy.[20] Likewise, at the same time in the United States, the *New York Times* featured letters and articles that not only described the plight of individual species—the Missouri beaver, the Labrador duck, the great auk, and of course the bison—but also summarized scientific understandings of the causes and evidence for prehistoric extinctions.[21] In every case, however, the concern or regret expressed was couched in romantic language that celebrated the beauty or utility of particular species, and did not reflect a more general concern with the preservation of "diversity" as such, or with what we might now consider the ecological or evolutionary consequences of extinction. What these writings do show, particularly in the case of the *New York Times* articles, is a public with a growing interest in and awareness of the lessons that scientists in fields such as paleontology had for understanding the economy of nature in the present.

In any event, it would be a mistake to associate the attitude of later nineteenth-century "conservationists" with those of conservation biologists today. Explorers and sportsmen like Theodore Roosevelt and Frederick Selous may have voiced concerns about the extermination of the charismatic animals they hunted, but these views did not necessarily translate into broader attitudes about diversity or concern for cultural conservation. Selous, an English explorer and personal friend of Roosevelt's, published a number of popular accounts of his travels, including *Sunshine and Storm in Rhodesia* (1896), where he discussed the theme of European expansion. He urged that "the whole question of the colonization by Europeans of countries previously inhabited by savage tribes must be looked at from a broad point of view," by which he meant that "final results" could justify sometimes unpleasant actions. As one example, Selous presented the "noble red man" who "has been exterminated by the more intelligent white man," but observed that "in place of a cruel, hopeless savagery there has arisen a civilization whose ideals are surely higher than those of the displaced barbarism" (Selous 1896, 65–66). Similarly, in South Africa "an orderly civilization has been established over a large area of this once savage country, and no one but an ignorant fanatic would, I think, assert that its present condition is not preferable from a humanitarian point of view to its former barbarism." While Selous's attitude was itself not exceptional for its time, his comments are particularly interesting because of the overt connection they drew to biological theory. As he concluded, "The British colonist is but the irresponsible atom employed in carrying out a preordained law—the law which has ruled upon this planet ever since, in the far off misty depths of time, organic life was first evolved upon the earth—the inexorable law which Darwin has aptly termed the 'Survival of the Fittest.'" (Selous 1896, 67). Selous's somewhat tenuous grasp of Darwin's own position is not the issue here (it was Spencer, not Darwin, who coined the term "survival of the fittest"). Rather, the point is that there is a clear connection between discussions of competition and extinction in elite scientific circles and the deployment of those ideas in popular and political discourse (such as in the earlier parliamentary discussions of the 1830s).

A very prominent nexus for biological and political discussions of racial extinction was the debate about the fate of freed slaves following the American Civil War. Historians are now beginning to draw attention to the calamitous health crisis that faced migrant African Americans during the Reconstruction era, and to the unpreparedness of the US federal government to cope with the problem.[22] But at the time, a number of white, mostly Southern physicians discussed the issue as an example of the unintended consequences of disturbing the system of slavery, couched in explicitly, though superficially, "Darwinian" terms. For example, in his essay "The Future of the Colored Race in the US," physician Eugene Rollin Corson argued that "it is to the school of Darwin, Wallace, and Spencer that we must turn" for guidance on the ultimate fate of the "negro" race, which, freed from the protective institution of slavery, must engage in a "struggle for existence" (Corson 1893, 197–98).

One of the most important discussions of this topic was Joseph Le Conte's book *The Race Problem in the South* (1892). Le Conte was formally trained as a physician, and he grew up in Georgia, serving in the Confederacy during the Civil War. But he also studied geology and natural history with Louis Agassiz at Harvard, and after the war joined the faculty in biology at the University of California. He published widely on topics relating to Darwinian thought, and his discussion of the "Negro problem" was explicitly couched in the language of evolution. In *The Race Problem*, Le Conte used an argument about extinction to justify the practice of slavery: Since Africans were by nature "inferior" to Europeans, the only alternative to slavery "would have been the extinction of the weaker race." He regarded the institution of slavery to be "a natural one," and insisted that "whatever is natural can not be wholly wrong" (Le Conte 1892, 354). The proof of this, for Le Conte, came in the "failure" of freed slaves to successfully compete with people of European descent (he conveniently ignored any political, sociological, or economic explanations). His more general conclusion was that

> In organic evolution the contact of two diverse forms determines either the extinction of the weaker or else its relegation to a subordinate place in the economy of Nature; the weaker is either destroyed or seeks safety by avoiding competition. In human evolution the same law must hold, with a difference to be determined by reason (Le Conte 1892, 359).

In essence, Le Conte argued that domination was preferable to extermination, although he also expressed the hope that African Americans had experienced sufficient "race evolution" to survive the transition from enslavement. However, he regarded the issue as one of broader import, since "everywhere the white race is pushing its way among the lower races. Everywhere, now that slavery is inadmissible, the result is gradual extinction of the lower race" (Le Conte 1892, 361–62). Whether "extermination, then, [was] the inexorable fate of all the lower races" Le Conte did not profess to know, but he speculated that the ultimate result of "the struggle for life and the survival of the fittest" among human races would be a final, perfect race that was most general and "coextensive with human nature" (Le Conte 1892, 375).

In the later nineteenth century, then, biological theories of competition and extinction could be deployed both as justification for European imperial expansion and as apologia for slavery. These attitudes began to shift somewhat during the first part of the twentieth century, but the first decade of that century did not see a sharp decline in the rhetoric surrounding inevitable racial extinction. At the same time, some authors began to question the logic of applying biological concepts of fitness and selection to human societies. For example, in 1902 the English economist John A. Hobson wrote a scathing critique of European imperialist ambitions in *Political Science Quarterly*, where he criticized the "dogmatism" of arguments that "defend the necessity, the utility, and even the righteousness of maintaining to the point of complete subjugation or extermination the physical struggle between races and types of civilization" (Hobson 1902, 460–61). He maintained that "imperialism is nothing but this natural-history doctrine [survival of the fittest] regarded from the standpoint of one's own nation," which he characterized as a belief that committed the naturalistic fallacy of equating "can" with "should" (Hobson 1902, 463). Yet Hobson ultimately adopted a Spencerian view of cultural competition, in which "we cease fighting with bullets in order to fight with ideas." In the end, "All the essentials of the biological struggle for life are retained—the incentive to individual vigor, the intensity of the struggle, the elimination of the unfit and the survival of the fittest" (Hobson 1902, 484). In other words, Hobson's main objection to "the imperialist argument" was that it relied overtly on violent force, and that it targeted entire groups rather than unfit individuals. He took for granted that humanitarianism that artificially propped up the "weak" in society caused more harm than good, and he concluded that "effective international government for national and racial selection can alone be regarded as an accurate and economical instrument of world progress" (Hobson 1902, 487).

Writing a year later, Lester Ward, the Yale sociologist and ardent Spencerian, had fewer qualms about violent extermination in his influential book *Pure Sociology* (1903). He opined that "war has been the chief and leading condition of human progress," and drew a direct analogy between "natural" and cultural extinction:

> In the organic world the struggle has the appearance of a struggle for existence. The weaker species go to the wall and the stronger persist. There is a constant elimination of the defective and a survival of the fittest. On the social plane it is the same, and weak races succumb in the struggle while strong races persist (Ward 1902, 184).

Likewise, in his paean to eugenics, the English statistician Karl Pearson unapologetically declared, "History shows me one way, and one way only, in which a high state of civilization has been produced, namely, the struggle of race with race, and the survival of the physically and mentally fitter race" (Pearson 1905, 21). Pearson argued that although the struggle between races was often "terrible," the outcome, a higher level of civilization, more than counterbalanced any suffering along the way; such struggle was "the fiery crucible out of which comes the finer metal" (Pearson 1905, 26). His arguments explicitly rejected the value of cultural diversity, for he derided "the romantic sympathy for the Red Indian generated by the novels of Cooper and the poems of Longfellow," and he stressed that "the nation organized for the struggle must be a *homogeneous* whole" (Pearson 1905, 25, 50). The broad picture Pearson painted was stark, and as redolent of Victorian cultural and biological attitudes as anything published thirty or forty years earlier:

> Mankind as a whole, like the individual man, advances through pain and suffering only. The path of progress is strewn with the wreck of nations; traces are everywhere to be seen of the hecatombs of inferior races, and of victims who found not the narrow way to the greater perfection. Yet these dead peoples are, in very truth, the stepping-stones on which mankind has arisen to the higher intellectual and deeper emotional life of today (Pearson 1905, 64).

Conclusion

The tenets of the dominant nineteenth-century view of extinction were: (1) Extinction is a regular, law-abiding, and natural process. (2) Extinction is driven primarily by competition; individuals or species that become extinct have failed to remain adapted to their environments, or have failed to compete for resources, and therefore "deserve" to die. (3) Extinction is inevitable; it is the logical consequence of natural selection. (4) Extinction tends to be equally balanced by the appearance of new species (speciation), thus maintaining the "economy of nature." (5) The number of taxa (i.e., the diversity) in the world therefore exists in dynamic equilibrium. The corollary to these tenets was the assumption that diversity was an inherent and self-renewing property of the "economy of nature," and thus required no special protection or independent valuation. As I have argued in this chapter, this particular way of conceiving of extinction was also implicated in a cultural and political imaginary—especially in Britain and the United States—that supported imperialism and downplayed the value of protecting species and peoples from threat of extinction. During the nineteenth century, at a time when naturalists understood nature to be an essentially endlessly renewable resource, diversity was taken for granted, and extinction was not perceived as a threat to the economy of nature. Therefore, diversity per se did not have normative value. Rather, extinction was understood to be nature's way of strengthening and improving itself by weeding out the unfit, and competition was celebrated as the source of natural progress. This view supported Victorian ideologies of social progress and imperial expansion, and justified a lack of concern about the inevitable victims of progress—combined with, at most, romantic nostalgia for cultures that passed away. When competition is natural, it was thought, extinction is inevitable and not to be resisted.

As I have suggested, both sets of views—about extinction, and also about the value of diversity—would begin to shift during the twentieth century. As was the case in the earlier period, the discourses surrounding extinction and diversity were enmeshed in a complex web involving biological, ecological, and anthropological theories, as well as political and cultural perceptions of nature and society. The next two chapters will begin to untangle these threads, focusing especially on new scientific perspectives that developed for the study of extinction dynamics and ecological systems, and which contributed to and reflected a new extinction imaginary dramatically different from the one it replaced.

3

CATASTROPHE AND MODERNITY

Ours is essentially a tragic age, so we refuse to take it tragically.
The cataclysm has happened, we are among the ruins, we
start to build up new little habitats, to have new little hopes.
It is rather hard work: there is now no smooth road into the
future: but we go round, or scramble over the obstacles.
We've got to live, no matter how many skies have fallen.

—D. H. Lawrence, *Lady Chatterley's Lover* (1928)[1]

With these words, the English writer D. H. Lawrence opened his controversial 1928 novel *Lady Chatterley's Lover,* a work remembered
mostly for its frank treatment of sexual themes that tested censorship
rules in its day. At first glance, it may seem an odd way to continue our
exploration of the history of extinction. Lawrence was not a biologist,
and he did not engage directly with biological or scientific themes in his
writing. But this passage captures the deep sense of pessimism, doubt,
and doom that pervaded European and American culture during the
decades between 1900 and the Second World War that marks a major
turning point in our story. Whereas the nineteenth century was characterized by a pervasive sense of optimism and faith in the potentially
limitless progress of Western civilization and its values, the early twentieth century presents us with a sudden and striking contrast. This is reflected in the arts, in philosophy, in social theory, in historical scholar-

ship, in political discourse, and in science as well: the positivism of the Victorian era was replaced by a far darker mood, in which to many contemporary observers the secure values of the previous society were called into question at seemingly every turn.

This new and pessimistic sensibility is sometimes simply called "Modernism," though as a general label the term fails to capture the specificity of the transformation I am describing. The eminent British historian Eric Hobsbawm has more aptly described the opening decades of the twentieth century as "the Age of Catastrophe," since the notion of "catastrophe" conveys the sense in which Western society, as Hobsbawm puts it, was experiencing a profound crisis "which, in one way or another, was in the process of destroying the bases of its existence, the systems of value, convention and intellectual understanding which structured and ordered it" (Hobsbawm 1989, 235). That sense of "catastrophe" or "crisis" is especially visible in the work of "literary Modernists" such as Lawrence, W. B. Yeats, T. S. Eliot, Ezra Pound, Louis-Ferdinand Céline, Virginia Woolf, and many others, where imagery of catastrophe or apocalypse often features prominently. But such themes also appear in contemporary philosophy—for example, in the work of Friedrich Nietzsche and Albert Camus; in the psychoanalytic theory of Sigmund Freud; in the pessimistic historical theory of Oswald Spengler; in the speculative fiction of H. G. Wells and Jack London; and in a variety of other cultural contexts, both "high" and "low." And this broad cultural sensibility—of crisis, catastrophe, and decline—did not fail to leave a mark on the science of the day. This was especially the case with theories of biological extinction, which tended to abandon the more progressive Darwinian account discussed in the last chapter in favor of explanations that invoked inevitable degeneration and "racial senility," often analogized directly with contemporary historical discussions of inevitable social and cultural decline and extinction.

One of the major themes in this book is the extent to which cultural and biological values surrounding extinction mirrored and reinforced one another, constituting what I am calling an "extinction imaginary." In the Victorian era, optimism about social and industrial progress resonated in biological theories emphasizing progress through healthy competition between organisms and species. Extinction was often seen as nature's way of clearing away dead brush to allow healthy roots to thrive. During the early decades of the twentieth century, however, growing pessimism about the very possibility of unlimited human progress found an echo in biological theories—popular especially among American and German paleontologists—that saw the extinction of species as the inevitable result of predetermined racial "life cycles," in which lengthy periods of "maturity" terminated in a final stage of racial "senility" or "senescence." In many cases, these biological theories explicitly invoked contemporary historical accounts—such as Spengler's—of similar cycles of social flourishing and decline which often forecast the imminent demise of Western civilization itself. The timing of the emergence of this new biological approach to understanding extinction is quite important: such theories, which fall under the general label of "orthogenesis," became popular between the 1880s and the 1910s, at precisely the time that pessimistic cyclical historical accounts came into vogue, especially in Germany. As I will show in this chapter, this was not mere coincidence; rather, it can be fairly conclusively demonstrated that such interpretations of human history directly influenced many of the leading proponents of cyclical biological theories.

Another important theme during this period is a climate of what might be called "apocalyptic thinking." This notion, exemplified in the Lawrence quotation that begins this chapter, held that Western society had reached such an advanced state of decay that only a dramatic and perhaps violent catastrophe could offer any hope of rebirth or renewal. Visions of apocalypse, either figurative or literal, feature prominently in poems and novels of the time, and often drew quite directly upon contemporary scientific concepts such as thermodynamics and extinction. Apocalyptic rhetoric was not incompatible with contemporary historical accounts of social decline, with one important distinction: whereas historians such as Spengler generally imagined repeating cycles of the rise, flourishing, and decline of civilizations, apocalyptic visions tended to emphasize a more linear conception of history, with final apocalypse representing a decisive culmination of some kind. In this sense, secular apocalypticism harked back to earlier Judeo-Christian roots in which a final apocalypse would mark the end of history. But while several prominent literary figures—Yeats, Lawrence, Eliot—often did flavor their visions of apocalypse with more traditional religious themes, a parallel strain of apocalyptic literature emerged, especially early science fiction, which imagined apocalypse in purely secular terms that did not necessarily hold out a promise of purification or redemption.

One might distinguish, then, between a kind of "secular millenarianism" in which a hoped-for cataclysm would redeem a corrupt society, and a more pessimistic apocalypticism in which humankind, at least, had little hope for a future. In the latter case, science fiction authors in particular imagined apocalyptic scenarios that invoked astronomical or geological events which strongly resembled earlier "catastrophic" geological theories, such as Cuvier's. Intriguingly, with a very few exceptions early-twentieth-century geologists and paleontologists shied away from overtly "catastrophist" accounts of extinction which, as the next chapter will discuss, reemerged only in the 1950s and 1960s. Here I will suggest that the secular apocalypticism of the pre–World War II era was an important cultural context for the later emergence of catastrophic mass extinction in scientific discourse—something that one might, in evolutionary terms, label a kind of "preadaptation" for the acceptance of catastrophic biological theories. Ultimately, however, the twentieth century presents a turning point in the Western extinction imaginary: increasingly, extinction was seen not only as an intrinsic check on the nineteenth-century dream of potentially limitless progress, but also as something of direct relevance and concern to the future of human civilization.

Degeneration and History

The roots of early-twentieth-century catastrophism can actually be traced back to the late Victorian era, when anxieties about social and biological "degeneration" became a prominent theme in European scientific and political discourse. As a concept, degeneration drew from fairly long-standing concerns about the future progress of racial and social development, mixing Darwinian and Spencerian evolutionary ideas with nineteenth-century theories of racial characteristics and contemporary pessimistic historiography. In a biological context, degeneration refers to the potential "reversion" of an individual or a species to a more "primitive type," or the failure of a race to maintain the "generative force" necessary to maintain adaptation or to progress to "higher" stages of evolution.[2]

Degeneration was also an important concern for early hereditarian biological theories: before the wide recognition of Mendel's laws of inheritance in the early twentieth century, there was much confusion and debate about how the mechanism of heredity functioned. In the first edition of *Origin of Species*, Darwin himself described mysterious cases in which "the child often reverts in certain characters to its grandfather or grandmother or other much more remote ancestor," though his theory of evolution via natural selection would appear to imply that such "reversions" would not become fixed in a population unless they conveyed immediate adaptive advantage.[3]

Darwin was mostly concerned here, and in his 1868 *The Variation of Animals and Plants under Domestication*, with cases where traits reappeared in humans or domesticated plants and animals after an absence of one or a few generations. But the discussion was complicated by ideas such as the German embryologist Ernst Haeckel's "recapitulation theory," which held that stages of fetal development of an animal (ontogeny) mirrored the evolutionary history of its lineage (phylogeny). If a developing fetus actually passed through earlier evolutionary stages, as Haeckel proposed, then it seemed possible that mature organisms themselves might have the capacity to revert to forms representing much earlier evolutionary steps. In fairness to nineteenth-century scientists, this is an enormously complex subject that still occupies the attention of modern biology, particularly in the field of evolutionary development (evo-devo). Modern genetics has identified mechanisms by which genes that become inactive can remain dormant in a genome for long periods of time, only to reappear through a rare mutation: so-called atavisms, such as the appearance of a vestigial tail in humans or legs in snakes. But the notion that a species or a race could "devolve," or the very idea of one species being more "primitive" than another, is not supported by modern evolutionary theory.[4]

Nonetheless, this is precisely what many late-nineteenth-century biologists believed, and the idea fit well with attempts to classify hierarchies of biological race in humans, for example, as discussed in the previous chapter. Furthermore, it occurred to some that allegedly "pathological" characteristics in humans—such as sexual deviancy, insanity, or criminality—could be explained as evolutionary degenerations. This was the argument infamously proposed by the Italian physician and criminologist Cesare Lombroso in his 1876 *L'uomo delinquente*, or *The Criminal Man*, which posited that criminals tend to exhibit physical characteristics that reflect those of primitive human ancestors or even apes.[5] Lombroso's work was deeply infused with assumptions of the now discredited science of anthropometry, which relied on skull measurements and other physical characteristics to infer emotional and mental capacities. But it was extremely influential in its time and was widely cited well into the twentieth century, especially in the United States by proponents of eugenics who opposed immigration from nations with high incidences of racial "degeneracy."[6] Degeneration also became a common trope in speculative literature, featuring prominently, for example, in works such as Edgar Rice Burroughs's *Tarzan of the Apes* (1914), H. G. Wells's *The Time Machine* (1895) and *The Island of Doctor Moreau* (1896), and Bram Stoker's *Dracula* (1897).

Degeneration is not a theory of biological or social extinction per se, but elements of degeneration theory found their way into a number of discussions of natural and human history that postulated a final stage of degenerate "senility" leading to the extinction of species or civilizations. One early and influential example was E. Ray Lankester's *Degeneration: A Chapter in Darwinism*, which was published in 1880. Lankester was a British zoologist who rose to prominence as a member of Thomas Henry Huxley's newly established University College London in the 1870s and 1880s, and he was a staunch ally of Huxley's who defended Darwin's mechanism of natural selection against challenges from Lamarckian and other non-Darwinian evolutionary theories. He defined biological degeneration as "a gradual change of the structure in which the organism becomes adapted to *less* varied and *less* complex conditions of life," which he based on extensive studies of the evolutionary histories of marine invertebrates.[7]

Degeneration thus led to a decrease in adaptive fitness that could ultimately result in the termination of a lineage; however, unlike many contemporary biologists and paleontologists who saw degeneration as the outcome of intrinsic organic life cycles, Lankester did not regard the degenerative phase as inevitable, and he situated the idea in an explicitly Darwinian context. He also drew analogies between biological and social degeneration, arguing that "high states of civilisation have decayed and given place to low and degenerate states," such as when "Rome degenerated when possessed of the riches of the ancient world" (Lankester 1880, 33, 58). In this sense Lankester, like other contemporary observers, connected social degeneration with excess of wealth or power that inhibited creativity, competition, and cultural progress. Of his own society he commented that among "the white races of Europe, the possibility of degeneration seems to be worth some consideration," and argued that "we have at least reason to fear that we may be degenerate" (Lankester 1880, 59–60).

Lankester's gloomy prognosis resonated strongly in a culture increasingly preoccupied with social and racial decline. Francis Galton, Darwin's first cousin and a widely acknowledged early proponent of eugenics, combined anxieties about race mixing, female emancipation, and cultural degeneration in an account of the fall of ancient Greek civilization that was a thinly veiled reference to Britain's own possible future. As he put it in his influential *Hereditary Genius* (1869),

> We know, and may guess something more, of the reason why this marvelously-gifted [Greek] race declined. Social morality grew exceedingly lax; marriage became unfashionable, and was avoided; many of the more ambitious women were avowed courtesans, and consequently infertile, and the mothers of the incoming population were of a heterogeneous [read: inferior] class. In a small sea-bordered country [much like England!], where emigration and immigration are constantly going on, and where the manners are as dissolute as were those of Greece in the period of which I speak, the purity of a race would necessarily fail. It can be, therefore, no surprise to us, though it has been a severe misfortune to humanity, that the High Athenian breed decayed and disappeared (Galton 1869, 344–45).

Indeed, degeneracy of this type became a great concern of the eugenics movement that emerged in the United States and Europe during the early twentieth century, where strategies designed both to strengthen the "pure" stock through selective breeding and to inhibit the influx of degenerate individuals through immigration restriction, forced sterilization, and (most horrifically) mass murder were pursued with state sponsorship. But one feature of the eugenics movement that distinguishes it from other similar biosocial approaches to degeneration — and a reason why it will not feature more prominently in this book — is its essential optimism: unlike many of the more pessimistic prophets of Modernism discussed in this chapter, most eugenicists believed firmly and idealistically in the power of modern science to redeem civilization and ensure its continued progress.[8]

In the study of human history, degeneration contributed to the amplification of an earlier rhetoric of inevitable social decay. While many mid- and later-nineteenth-century historians were swept up in the spirit of progress and Positivism that buoyed Victorian era imperialism, a notable faction both in Germany and Britain had taken a less optimistic stance. In the early nineteenth century, for example, the German historian Barthold Georg Niebuhr suggested disturbing parallels between the decline and fall of Rome and the future of his own contemporary civilization (and Niebuhr was one of the historians who influenced Charles Lyell's historical conception of earth history as cyclical rather than progressive). The French theorist August Comte's philosophy of Positivism — which identified cycles of social progress through theological, metaphysical, and scientific stages — was in many ways a direct response to such thinking, with the important distinction that Comte did not envision a period of decline that followed a civilization's ascendancy. But the gloomier tradition persisted, especially in Germany, where the work of Jacob Burkhardt influenced many with its pessimistic predictions about the future of contemporary European culture in the face of an encroaching "universal barbarism." Burckhardt's major works, published largely in the 1860s and 1870s, influenced a generation of historians; and through his scholarship and personal friendship, Burkhardt strongly influenced the development of Friedrich Nietzsche's own uniquely pessimistic philosophy.[9]

Probably the most concise and influential single account of degeneration as a historical force, though, was an 1892 book by Max Nordau titled simply *Degeneration*, which encapsulated most of the anxieties that would surface in various guises in the catastrophic Modernism of the next century. Nordau was born to an Austrian Jewish family, and he eventually became a major figure, along with Theodor Herzl, in the World Zionist Organization after concluding that Jewish emancipation would not succeed in Europe following the notorious Dreyfus affair in France in 1894. This gives Nordau's approach to degeneration a somewhat unique flavor, in that he was not, like many contemporary observers, defending traditional Nordic Christian European values. Nonetheless, he did diagnose in his own society a decline in morality and idealism, which he described as "the end of an established order, which for thousands of years has satisfied logic, fettered depravity, and in every art measured something of beauty" (Nordau 1895, 5).

Nordau took exception to the growing popularity of the notion of fin-de-siècle (end of the century), since he regarded as absurd the idea that artificial units of time like "century" have independent historical meaning. Rather, he proposed the term *fin-de-race* as more accurately capturing "the prevalent feeling . . . of imminent perdition and extinction" felt across Europe. "In our days," he wrote, "there have arisen in more highly-developed minds vague qualms of a Dusk of the Nations, in which all suns and all stars are gradually waning, and mankind with all its institutions and creations is perishing in the midst of a dying world" (Nordau 1895, 2). He justified this overtly catastrophic vision by citing studies of biological degeneration—including Lombroso's in particular—as contributing to an increasingly unhealthy society that would be incapable of perpetuating itself: "Degeneracy is a pathological state; the most convincing proof of this is, that the degenerate type does not propagate itself, but becomes extinct" (Nordau 1895, 555). In addition to the now familiar references to social and cultural decay, Nordau also drew metaphors from natural catastrophe, including the famous eruption of the Indonesian volcanic island of Krakatoa in 1883 that killed tens of thousands and disrupted global weather patterns for years: "Massed in the sky the clouds are aflame in the weirdly beautiful glow which was observed for the space of years after the eruption of Krakatoa. Over the earth the shadows creep with deepening gloom, wrapping all objects in a mysterious dimness. . . . The day is over, the night draws on" (Nordau 1895, 6).

Without doubt, the great traumatic event of the early twentieth century was the First World War, which profoundly influenced the dismal mood of European literature, philosophy, and history during the 1920s and 1930s. This was the "cataclysm" Lawrence referred to in the opening of *Lady Chatterley's Lover*, which had "brought the roof down over" Constance Chatterley's head: Having been brought back from Flanders to England "more or less in bits," Constance's husband Clifford attempted to resume normal life, "but he had been so much hurt that something inside of him had perished, some of his feelings were gone" (Lawrence 1928, 1–2). Constance's later affair, then, is traceable directly back to this calamitous event. But while it is impossible to exaggerate the trauma inflicted on the European psyche by the war, it is important to emphasize that, as the preceding discussion has indicated, the war did not create the culture of doom and catastrophe that permeated the first half of the twentieth century. On the contrary, to many observers it simply confirmed the dire predictions they had been making for some time. One important effect war may have had, however, was to enhance pessimism towards redemptive hopes that secular millenarians clung to with increasing tenuousness—though, as I will discuss in the next chapter, the final blow probably did not come until the Second World War.

Oswald Spengler's great opus *The Decline of the West*—a monumental historical treatise on the rise and fall of the world's civilizations—is often seen to epitomize the mood of postwar Europe, and indeed it was received that way. As the German philosopher Ernst Cassirer described it in 1946,

> In 1918 there appeared Oswald Spengler's *Decline of the West*. Perhaps never before had a philosophical book such a sensational success. It was translated into almost every language and read by all sorts of readers—philosophers and scientists, historians and politicians, students and scholars, tradesmen and the man in the street.

But as Cassirer went on to make clear, Spengler's book galvanized feelings that had been deeply felt for some time:

> The title *Der Untergang des Abendlandes* was an electric spark that set the imagination of Spengler's readers aflame. The book was published in July, 1918, at the end of the first World War. At this time many, if not most of us, had realized that something was rotten in the state of our highly praised Western civilization. Spengler's book expressed, in a sharp and trenchant way, this general uneasiness (Cassirer 1946, 289).

While *Decline of the West* was not published until after the armistice in 1918, it had in fact been conceived as early as 1911, shortly after a modest inheritance following his mother's death allowed Spengler to resign a position as a high school teacher and to pursue scholarship full-time. As Spengler himself explained it, "At that time the World-War appeared to me both as imminent and also as the inevitable outward manifestation of the historical crisis, and my endeavour was to comprehend it from an examination of the spirit of the preceding centuries—not years" (Spengler 1926, 46). In Spengler's own formulation, then, the work itself—and its theory of cycles of rise and fall of civilizations—was less inspired by the war as the war was "the inevitable outward manifestation" of the theory, which had much deeper antecedents. In particular, Spengler cited Nietzsche and Johann Wolfgang von Goethe as his major inspirations, and his view of history owed a deep debt to the nineteenth-century tradition of Romantic and pessimistic historicism.

The main thesis of *Decline of the West* is that civilizations, like organisms or species, have predetermined life cycles that last no more than about a thousand years. As Spengler put it, "Every Culture, every adolescence and maturing and decay of a Culture, every one of its intrinsically necessary stages and periods, has a definite duration, always the same, always recurring with the emphasis of a symbol" (Spengler 1926, 109). To prove this point, the book surveys the great civilizations of "world history," and argues that the Western belief in its own exceptionalism is analogous to the "geocentrism" of cosmology before Copernicus. Consequently, Spengler described his viewpoint as "Copernican," since "it admits no sort of privileged position to the Classical or the Western Culture" (Spengler 1926, 18). Rather, as his work shows, world history is "the drama of *a number* of mighty cultures, each springing with primitive strength from the soil of a mother-region to which it remains firmly bound throughout its whole life-cycle . . . each having *its own* idea, *its own* passions, *its own* life, will, and feeling, *its own* death" (Spengler 1926, 21). On this last point, Spengler was quite explicit that the "death" of a civilization is analogous to the biological extinction of species, since "each Culture, further, has *its own mode of spiritual extinction*, which is that which follows of necessity from its life as a whole" (Spengler 1926, 356).

What is remarkable about Spengler's theory of history—and this has, I think, been missed by many readers—is how much it engages with biology and geology. Spengler's philosophy was broadly anti-Darwinian and Romantic, which is to say he rejected the strict materialism of natural selection and random genetic mutation as the basis for evolutionary change. Indeed, his belief that human civilizations have "life cycles" is founded on an organic conception of social organization that owes much to the "vitalist" tradition in Romantic German biology of the nineteenth century: the notion that, as he put it, to each individual organism or species "is given also a definite *energy* of the form—by virtue of which in the course of its self-fulfillment it keeps itself pure or, on the contrary, becomes dull and unclear or evasively splits into numerous varieties" (Spengler 1926, 32). Spengler's use of the term "form" here echoes Goethe's use of the term "morphology" (the term Spengler uses in the German original is *morphologie*), which Goethe frequently employed to discuss anatomical similarities between organs of distinct species of organisms (what we would now call "homologies"—for instance, the analogy between the wing of a bat and the fin of a dolphin). Spengler paid this debt to Goethe in the subtitle to the German edition of his book, *Umrisse einer Morphologie der Weltgeschichte* (Outlines of a Morphology of World History), which refers to the analogies Spengler detected between the stages of the various civilizations he studied.

In Spengler's eyes, the mechanism that explains the development of organisms, species, or civilizations through distinct and predetermined stages of a life cycle is an intrinsic property—a vital "force"—that cannot be reduced to what Spengler saw as lifeless mechanical causes. Spengler viewed Lyell's and Darwin's accounts of natural history as "but derivatives of the development of England itself," which "in place of the incalculable catastrophes and metamorphoses such as von Buch and Cuvier admitted . . . put a methodical evolution over very long periods of time and recognize as Causes only *scientifically calculable* and indeed *mechanical utility-causes*" (Spengler 1926, 31). But Spengler had reason to question this mechanical account, since "there is no more conclusive refutation of Darwinism than palaeontology." Rather than the smooth unbroken series of transitional forms Darwin predicted the fossil record should reveal, Spengler argued, "we find perfectly stable and unaltered forms persevering through long ages, forms that have not developed themselves on the fitness principle, but *appear suddenly and at once in their definitive shape*; they do not thereafter evolve towards better adaptation, but become rarer and finally disappear, while quite different forms crop up again" (Spengler 1926, 32). The only explanation for this, Spengler believed, was an internal vital force responsible for the birth, flourishing, and death of species, which he described as "*a life-duration of this form*, which . . . leads naturally to a senility of the species and finally to its disappearance" (Spengler 1926, 32). And he was absolutely clear that it was this same "life-duration of form" that explains

> the swift and deep changes [that] exert themselves in the history of the great Cultures, without assignable causes, influences, or purposes of any kind. The Gothic and the Pyramid styles come into full being as suddenly as do the Chinese imperialism of Shi-hwang-ti and the Roman of Augustus, as Hellenism and Buddhism and Islam (Spengler 1926, 33).

It is important to note that while Spengler's biological analogy may seem out of sync with our current understanding of evolutionary biology, it was quite widely supported (as will be discussed below) by a number of influential contemporary German and American paleontologists, who also advocated theories of species life cycles. Spengler was obviously well aware of these theories, and even predicted that "without doubt the biology of the future will—in opposition to Darwinism and to the exclusion in principle of causal fitness-motives for the origins of species—take these *preordained* life durations as the starting point for a new enunciation of its problem" (Spengler 1926, 109). Spengler was wrong in this prediction—he did not foresee the Modern Evolutionary Synthesis of the 1940s, which with ample reason would establish Darwinism as the unchallenged model for evolutionary explanation. But in its time, Spengler's work was a comprehensive system of human history permeated *through and through* with biological thinking that was up to date with regard to explanations, in particular, for extinction. Furthermore, as I will discuss shortly, this influence was not one-way; paleontologists of the era were as likely to invoke analogies between natural and human history as were social theorists, and a number of influential German paleontologists took explicit inspiration from Spengler's work in developing their own theories of intrinsic species life cycles.

Apocalypticism, Cataclysm, and Modernism

Depictions of cataclysmic geological upheaval would become common as metaphors in Modernist literature of the next several decades, but they would also feature more literally in speculative apocalyptic science fiction, as well as in ostensibly scientific accounts of natural and human history. Catastrophe was also a prominent theme in the "cataclysmic" history and social theory of a group of late nineteenth- and early-twentieth-century American intellectuals that included Ignatius Donnelly, Homer Lea, Brooks and Henry Adams (direct descendants of US Presidents John and John Quincy Adams), and Jack London (the famous adventure novelist). While this group was quite heterogeneous in background, belief, and genre of expression—ranging from the speculative popular geology of Donnelly to the highbrow economic histories of the Adams brothers and the pulp fiction of London—it can be loosely characterized by a shared belief that society had reached a dangerous impasse, threatened both from within by immigration, labor unrest, and predatory capitalism, and from outside, especially by inevitable racial conflict with an ascendant Japan. Although biological degeneration did not figure prominently in the works of these authors, American "cataclysmists" (as the historian Frederic Jaher has labeled them) shared with their European counterparts a foreboding that "a bleak future in which sudden destruction or slow stagnation lay in wait for a rotten civilization" (Jaher 1964, 7).

For some cataclysmists, catastrophe could take the form of a violent physical event or bloody global war—as was predicted by Homer Lea, whose *Valor of Ignorance* (1909) imagined a catastrophic war between the United States and Japan. For others, including Brooks and Henry Adams, the catastrophe was more likely to be economic or political. This sense was probably amplified by the great economic uncertainly in the United States during the decades between the 1850s and the 1890s, where cycles of panic and economic depression severely destabilized the traditional social structures that the Adamses, in particular, as Boston "Brahmins," had taken for granted. A major economic depression in the 1870s may have catalyzed this thinking, as did militant labor organization and violent strikes throughout the 1880s. These included the notorious Haymarket Affair in Chicago, in which a group of "anarchists" was convicted of inciting a riot in which several police officers were killed; the Homestead Strike of 1892, in which Pennsylvania steel workers clashed violently with Pinkerton agents employed by Andrew Carnegie; and the Panic of 1893, on the eve of the World's Columbian Exposition in Chicago, that saw nationwide unemployment rates reach as high as 18 percent.

The reaction of the Adams brothers is best exemplified in Brooks Adams's 1895 book *The Law of Civilization and Decay*, which fits within the tradition of pessimistic cyclical historiography of Burkhardt and later Spengler, but has a uniquely "American" focus in its obsession with rampant capitalism, which Adams blamed for most of the social disintegration plaguing modern society. Adams presented his theory as scientific, since he based his "law" of civilization on "the accepted scientific principle that the law of force and energy is of universal application in nature, and that animal life is one of the outlets through which solar energy is dissipated" (Adams 1896, viii–ix). His theory, then, roughly proposed that a kind of thermodynamics applies to human societies, where civilizations have natural endowments of "energy" that are eventually dissipated through war or scientific and industrial production, leading either to stagnation or reversion to a more "primitive form of organism" (Adams 1896, xi). Needless to say, he viewed his own contemporary society as exhibiting symptoms of extreme dissipation, mostly thanks to the centralization of capital by powerful—and Jewish—families such as the Rothschilds (Adams was an overt anti-Semite), and he predicted that it must eventually give way to some new, more energetic, civilization. In this regard, Adams found nothing special in the fate of Western culture, since his historical survey showed "a progressive law of civilization, each stage of progress being marked by certain intellectual, moral, and physical changes" that would repeat through endless cycles of struggle, consolidation of energy, and dissipation. (Adams 1896, 362).

In contrast to the blue-blooded Adams, Ignatius Donnelly was a populist agrarian with utopian leanings whose early career was spent in politics, where he represented Minnesota in the US Congress during the 1860s before retiring to private law practice and amateur scientific writing. After leaving politics Donnelly channeled these interests into pseudoscientific works in which he proposed cycles of astronomical and geological catastrophe that he alleged had shaped human history. The first of those works, *Atlantis: The Antediluvian World* (1882) was a popular success and inspired a revival of interest in the Atlantis myth that persisted well into the later twentieth century. The second was a more ambitious treatise, grandiosely entitled *Ragnarok: The Age of Fire and Gravel* (1883), which proposed to explain major changes in human history as the result of astronomical cataclysms in the distant past and perhaps future.

Ragnarok made its debt to geological catastrophism clear on its title page, where it quoted Cuvier's statement in his "Preliminary Discourse" to *Revolutions on the Surface of the Globe*:

> I am not inclined to conclude that man had no existence at all before the great revolutions of the earth. He might have inhabited certain districts of no great extent, whence after these terrible events he re-peopled the

world. Perhaps, also the spots where he inhabited where swallowed up and his bones lie buried under the beds of the present seas.[10]

Indeed, Donnelly's central argument was that long ago—but well within the bounds of human prehistory—a comet had impacted the earth, setting off a dramatic series of cataclysms that included vaporized seas and huge storms, earthquakes and volcanic eruptions, and ultimately a global ice age which lowered temperatures so dramatically that its effects are still felt in the present (fig. 3.1). Donnelly's "evidence" for this event was what he referred to as "the Drift": a layer of sediment at the top of the earth's crust containing clay, gravel, fragmented stones, and evidence of recent human and animal life, but no fossils, which he argued could only be "the result of violent action of some kind" (Donnelly 1883, 7). Further evidence of this catastrophe, he argued, is found in the mythologies of civilizations from ancient Egypt to classical Greece to Judaism and Islam: why else, he wondered, would nearly every society record ancient cataclysms, like the flood stories that appear so frequently in myths and scriptures from around the world? Donnelly proposed that the profound cataclysm that created the Drift had occurred perhaps as long as thirty thousand years ago, and wiped out a thriving civilization—perhaps the one described in the Atlantis myth. But he left open the possibility that other, similar astronomical and geological catastrophes had struck at other points in human history, and even that such an event might occur again, as part of a grand cycle of destruction and rebirth: "In endless series the ages stretch along—birth, life, development, destruction. And so shall it be till time is no more" (Donnelly 1883, 436). But, perhaps influenced by his devout Catholic upbringing, Donnelly held out hope that our current civilization might be spared by divine intervention, provided that society proves itself to be virtuous and worthy. "From such a world," he wrote, "God will fend off the comets with his great right arm, and the angels will exult over it in heaven" (Donnelly 1883, 441).

While it was presented as a scientific theory, one would be hard pressed to distinguish many elements of Donnelly's comet impact hypothesis from contemporary speculative fiction, which around 1895

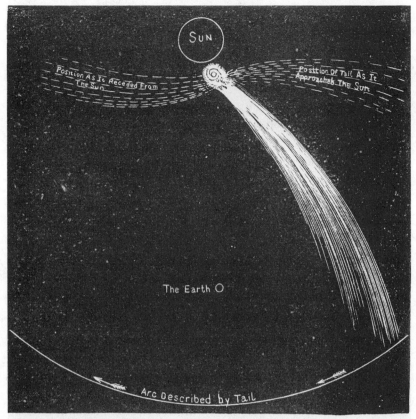

THE COMET SWEEPING PAST THE EARTH.

FIGURE 3.1 A depiction of a comet crossing the path of the earth, in Ignatius Donnelly, *Ragnarok: The Age of Fire and Gravel* (New York: Appleton and Co., 1883), 91. As Donnelly speculated in *Ragnarok*, the earth may have passed through the tail of a mighty comet. "Or, on the other hand, the comet may, as described in some of the legends, have struck the earth, head on, amid-ships, and the shock may have changed the angle of inclination of the earth's axis, and thus have modified permanently the climate of our globe" (Donnelly 1883, 94–95).

began to demonstrate increasingly apocalyptic preoccupations. Indeed, it might be fair to say that the genre of "apocalyptic" fiction was invented in this period. The idea of apocalypse is of course a very old one, and in the Christian tradition it extends back many centuries. But I am distinguishing Modernist "secular apocalypticism" from the more explicitly religious variety for several reasons. First, Christian apocalyptic thinking tends to view the apocalypse as a purifying and redemptive event, whereas the secular version is far less optimistic about a new beginning for humanity after the catastrophe. Second, in religious traditions humans occupy the central focus of the historical stage, so that the apocalypse that ends the human drama is also literally "the end of the world." In contrast, by the late nineteenth century secular apocalyptic visions often presented the end of humanity as simply an event in the continuation of a broader natural history; many apocalyptic novels and stories imagined plants and animals reclaiming a world vacated by human beings. This attitude was no doubt influenced by the nineteenth-century discovery of "deep time," which, as Martin Rudwick has argued, was a revolution in human awareness no less profound, or potentially unsettling to notions of human importance, than the Copernican one.[11] Finally, the agent of apocalyptic catastrophe in the secular context is always some kind of naturalistic event, whether external to human affairs (e.g., a comet, plague, or other natural disaster) or internal (war or an industrial or scientific accident). In this regard, apocalyptic fiction of the early twentieth century drew quite explicitly on contemporary science to imagine realistic scenarios, and tended to avoid arbitrary supernatural agents of destruction.

The fascinating question to ask is why apocalyptic literature appeared so suddenly, around the turn of the twentieth century, and with almost no precursors. Based on a fairly careful census, I find that the period between 1800 and 1895 saw the publication of only a small handful of stories, poems, and novels imagining a secular apocalypse; between 1895 and 1945 the number increased to several dozen; and after 1945 apocalyptic novels, stories, and films number in the hundreds. I argue that this phenomenon is a product of the same shift in thinking about Western progress—cultural and scientific—that characterized anxieties about degeneration and decline in the other manifestations we have already discussed. In fact, it would not be too strong to say that the very idea of the end of humanity, in a secular context, only became thinkable in the context of these broader anxieties, and that this apocalyptic "imaginary" helped to create a context in which scientific theories of catastrophic extinction would come to be taken more seriously.

Mary Shelley's speculative novel *The Last Man*, published in 1826, is often considered the first modern apocalyptic novel (the French novel *Le dernier homme*, published in 1805, is sometimes given that distinction, but because it has strong religious elements I do not consider it properly "secular"). Shelley's novel, which was written shortly after the death of her husband, the great Romantic poet Percy Bysshe Shelley, tells the story of the sole human survivor of a worldwide plague sometime in the twenty-first century. While the novel can be read as a critique of science in the same vein as her earlier and much more successful *Frankenstein* (1818), the overwhelming theme of *The Last Man* is rather the intense loneliness and isolation of the main character. As a number of scholars have suggested, this mood was inspired by Shelley's own feeling of isolation after the deaths of her husband and Lord Byron, and the consequent unraveling of their circle of Romantic idealists.[12] What is especially noteworthy about *The Last Man*, however, is how poorly it was received. This owes in part to the quality of the narrative—both the prose and the plot are well below the standard of *Frankenstein*—but there is equal reason to suspect that the subject itself was unpalatable to contemporary tastes. The reviews were scathing: one reviewer described it as "a sickening repetition of horrors," and another as "the offspring of a diseased imagination, and of a most polluted taste," while a third simply called it an "abortion."[13] Nor was the reading public very interested: it sold the least well of all of Shelley's novels, and drifted into obscurity in the decades after its publication.

The Last Man was not the only such work of its time, but it was the longest and most ambitious. George Gordon, Lord Byron's poem "Darkness," published in 1816, also imagined a civilization-ending catastrophe, but this event was couched in fairly oblique and metaphorical terms. The poem imagined a future in which

> The bright sun was extinguish'd, and the stars
> Did wander darkling in the eternal space,
> Rayless, and pathless, and the icy earth
> Swung blind and blackening in the moonless air;

and presented the dismal prospect where

> The world was void,
> The populous and the powerful was a lump,
> Seasonless, herbless, treeless, manless, lifeless —
> A lump of death — a chaos of hard clay (Byron 1816).

Not only was the message of the poem quite bleak, offering little in the way of hope for humankind's salvation, but the imagery itself was drawn from contemporary events: in 1816 the Indonesian volcano Mount Tambora erupted, darkening skies worldwide with ash and causing the "Year without a Summer" in Europe, which Byron claimed was the initial inspiration for his poem.

Not to be outdone, the Scottish poet Thomas Campbell published a poem titled "The Last Man" in 1823, which occasioned a minor priority dispute with Byron (some observers commented on the similarities between Campbell's and Byron's imagery, leading Campbell to claim that he had in fact suggested the idea to Byron before the publication of "Darkness"). But while Byron's poem held out little hope of redemption for humankind (the final lines read "And the clouds perish'd; Darkness had no need / Of aid from them — She was the Universe"), Campbell was explicit in presenting apocalypse as a precursor to divine redemption:

> This spirit shall return to Him
> Who gave its heavenly spark;
> Yet think not, Sun, it shall be dim
> When thou thyself are dark!
> No! it shall live again, and shine
> In bliss unknown to beams of thine,
> By Him recalled to breath,
> Who captive led captivity,
> Who robbed the grave of victory,
> And took the sting from death (Campbell 1823).

In this way, while the two poems share many formal similarities, their messages were quite different: as one literary scholar has pointed out,

Byron's poem was among the first works of literature that "envisaged apocalypse without millennium," something it shared in common with Shelley's novel, the collective weight of which "moved almost the entire critical establishment to deny the possibility of imagining Lastness" (Paley 1993, 109).

However unwilling Georgian and Victorian readers may have been to imagine "lastness," by the end of the nineteenth century things had changed significantly. Between 1885 and the Second World War a raft of stories and novels depicting the extinction or near extirpation of the human race were published, many of which invoked astronomical or geological catastrophes. One of the first such works was Richard Jefferies' novel *After London* (1885), which described the aftermath of an unnamed catastrophe in which human survivors revert to a pastoral, agrarian lifestyle reminiscent of the Middle Ages. Jefferies was a nature writer with decidedly Romantic leanings, so it is not surprising that he approved of his fictional development; in his disdain for industrialized society, his views closely match those of contemporary agrarian apocalypticists in the United States. But the writer probably most closely associated with early apocalyptic science fiction was the English socialist writer and social critic H. G. Wells, whose novels and stories explored themes of catastrophe, degeneration, and apocalypse and reached an enormous reading public.

Wells' best known apocalyptic novel is *The Time Machine* (1895), a "scientific romance" in which a contemporary English time traveler visits a distant future in which humanity has split into two distinct races: the peaceful, childlike Eloi, whom he discovers to be the food supply for the subterranean, brutish Morlocks. After escaping from the Morlocks, the traveler visits an even more distant future in which nearly all life on earth has been extinguished, and the sun hangs reddened and dying above a bleak landscape: "Beyond the lifeless sands the world was silent—silent! It would be hard to convey to you the stillness of it. All the sounds of man, the bleating of sheep, the cries of birds, the hum of insects, the stir that makes the background of our lives, were over" (Wells 1895, 201). The book thus manages to imagine three distinct types of extinction: the "racial senescence" of humanity, the extinction of life on earth, and the heat death of the sun itself.

It was no accident that Wells's visions of apocalypse drew heavily on contemporary scientific theories and anxieties. Despite having had a rather eclectic formal education, Wells studied biology with T. H. Huxley and was intimately familiar with Lankester's theory of biological degeneration, which informed his characterizations both of the senile, ineffectual Eloi (essentially the European upper class) and the degenerate Morlocks, who represented the brutish lower classes. These were themes Wells also explored a year later in *The Island of Doctor Moreau*, which describes the discovery of a mysterious island where an insane scientist is experimenting with creating animal-human hybrids. Despite the monstrousness of these artificial hybrids, the protagonist's experience among them ultimately causes him, upon his return to civilization, to see his fellow humans as in the process of degenerating to an animal state.

In a short story from this period, "The Star" (1897), Wells imagined a different kind of extinction, where the world is threatened by the discovery of a strange celestial "wanderer" on a trajectory for a seemingly inevitable catastrophic head-on impact with the earth. While the impending event sets off mass panic and hysteria, it is a near miss in the end: though much of humanity is killed by earthquakes, tsunamis, and volcanoes caused by the gravitational disruption of the wanderer, humanity survives and is even inspired towards a "new brotherhood" in the aftermath. Finally, in his 1914 novel *The World Set Free*, Wells conjured a new kind of human-instigated catastrophe, in which the power of radioactivity (recently discovered, in 1896, by Henri Becquerel and identified and named by Marie and Pierre Curie) is harnessed to produce devastating, continual explosions that are never fully exhausted. Although his prediction about the exact nature of eventual nuclear explosives was inaccurate, Wells was quite familiar with the latest atomic science, and may even have read the physicist Frederick Soddy's much-publicized 1903 comment that knowledge of radioactivity must "make us regard the planet on which we live rather as a storehouse stuffed with explosives, inconceivably more powerful than any we know of, and possibly only awaiting a suitable disaster to cause the earth to revert to chaos" (Soddy 1903, 720).

While Wells was certainly the most popular author of apocalyptic fiction at the time, he was far from the only one. Between 1900 and 1939 a number of similar works were published, such as M. P. Shiel's *The Purple Cloud* (1901), Frank Lillie Pollock's "Finis" (1906), E. M. Forster's "The Machine Stops" (1909), Jack London's *The Scarlet Plague* (1912), William Hope Hodgson's *The Night Land* (1912), George Allen England's *Darkness and Dawn* (1914), S. Fowler Wright's *Deluge and Dawn* (1928), Wells's *The Shape of Things to Come* (1933), Stanley G. Weinbaum's *The Black Flame* (1934), and many others. Collectively, these stories imagined apocalypse coming in the form of plague; poison gas; solar extinction; extraterrestrial impact; flood; and, increasingly after 1917, massive war. Unlike Wells's descriptions, which were in many ways rather tame, these novels and stories often described the consequences of catastrophe in gruesome detail, lingering on the horror experienced by the survivors and the often grisly effects of the catastrophe. Clearly, there was a growing reading public with a fascination for such stories, and with a tolerance for what would have been considered "the offspring of a most diseased imagination" during the previous century. This no doubt reflects changing social mores and tolerances in Europe and the United States resulting from public exposure — through journalism, photography, and eventually film — to horrific scenes of war and natural disaster in the later decades of the nineteenth century and early twentieth. (In the United States, for example, the Civil War photographs of Mathew Brady, first displayed in the 1860s, were the first unfiltered view of the carnage of war to which the public was exposed.) But something else is required to explain the avid fascination that writers and readers had developed for imagining the end of the world in increasingly detailed and scientifically accurate terms — a trend that has continued to the present day, as will be discussed in later chapters.

In some of this literature, the influence of contemporary theories of biological and social degeneration or cyclism is clear. In addition to Wells, Jack London also invoked themes of degeneration, particularly in his novel *The Scarlet Plague*, which described the aftermath of a worldwide plague (in 2013) that decimated the human population and reduced the survivors to bitter and animalistic struggle. In recounting the immediate devastation that followed the plague, London explained: "In the midst of our civilization, down in our slums and labor-ghettos, we had bred a race of barbarians, of savages; and now, in the time of our calamity, they turned upon us like the wild beasts they were and destroyed us" (London 1915, 30). In London's pessimistic vision, it was precisely the most brutal and unrefined who were most successful following "the calamity," just as it was the "weeds and wild bushes" that survived while "soft and tender" domesticated crops were wiped out. London also invoked a cyclical view of human history in which "the human race is doomed to sink back farther and farther into the primitive night ere again it begins its bloody climb upward to civilization," despite the ultimate truth that "just as the old civilization passed, so will the new. . . . All things pass" (London 1915, 12, 52).

M. P. Shiel's *The Purple Cloud*, a story about a catastrophe caused by the volcanic release of poison gas, took a different scientific basis—the "catastrophist" geology of Cuvier—in presenting an updated "last man" narrative. In this case, following a geological catastrophe, a single human survivor travels an empty world, witnessing scenes of death and horror everywhere he turns, eventually going insane and declaring himself emperor of the world. In describing his plunge into megalomaniacal madness, the narrator of the novel meditates on his own descent from the "Western, 'modern' mind" to "a primitive and Eastern one" (Shiel 1901, 87). Eventually he encounters a young woman, and after considering "the nobility of self-extinction," he opts to restart the human race.[14] In addition to being fascinated with theories of geological catastrophe, Shiel, an Englishman of mixed Irish and West Indian ancestry, was also strongly influenced by late Victorian theories of racial degeneration.

Ultimately, early apocalyptic fiction was part of the same milieu in which social and biological degeneration, evolutionary theory and geology, contemporary physics, and other "scientific" themes met anxieties about social, economic, and political disruption to form a powerful cultural discourse. Readers of pulp novels were exposed to this culture of pessimism just as were readers of "highbrow" literature: the prose may have been less artful, but the descriptions of catastrophe and apocalypse were no less vivid in the work of Wells or Shiel than, say, in the poetry of Yeats or Eliot. Like Lawrence's opening to *Lady Chatterley*, some of the most famous lines from Modernist literature of this period are vivid with apocalyptic despair. In poetry, Yeats's "The Second

Coming" described the terror of a society in which "things fall apart; the centre cannot hold," and "mere anarchy is loosed upon the world," while Eliot conjured a vision of "hooded hordes swarming / Over endless plains" in "The Waste Land" (1922), and pictured a "valley of dying stars" before declaiming, "This is the way the world ends / Not with a bang but a whimper" in "The Hollow Men" (1925). Likewise, in prose fiction Joseph Conrad's narrator in *Heart of Darkness* (1899) describes a London sunset as "a dull red without rays and without heat, as if about to go out suddenly, stricken to death by the touch of that gloom brooding over a crowd of men." William Faulkner's *The Sound and the Fury* (1929) recalls the gift of a watch from father to son as "the mausoleum of all hope and desire," since time "only reveals to man his own folly and despair." And Stephen Dedalus, the protagonist of James Joyce's *Ulysses* (1922), describes history as "a nightmare from which I am trying to awake." It is noteworthy, too, that many of these novels and poems eschew traditional structure and narrative resolution. Novels by Faulkner, Joyce, Woolf and others, for example, employ stream-of-consciousness narrative that emphasizes the incoherence or irresolvability of human experience—a condition aptly encapsulated in Faulkner's title allusion to Macbeth's soliloquy in which life "is a tale told by an idiot, full of sound and fury, signifying nothing."

History, Biology, and Extinction

Up to this point, we have considered historical cyclism, degeneration, and extinction from the standpoint of broader cultural attitudes. It is now time to examine these topics from a scientific perspective. The broad argument I will make, though, is that, as with all extinction imaginaries, it is impossible to neatly distinguish "science" and "culture" when it comes to discussing these themes during the early part of the twentieth century. As has already been suggested, science played a major role in the literary imagination of decline and apocalypse at the turn of the century, and accounts of the rise and fall of civilizations drew frequent and direct analogies with organic theories of evolution and degeneration. Likewise, as we will see, scientific understandings of extinction and cyclical organic development were often explicitly linked to contemporary understanding of human history and progress, and many paleontologists and biologists were unwilling to draw a clear line between "laws" of human and natural historical development.

The decades between 1880 and 1940 were a complicated and confusing time in evolutionary biology. Sometimes referred to as the "eclipse of Darwin," this period saw a proliferation of evolutionary theories—from old-fashioned Lamarckism to newer ideas, such as "mutationism," based on early Mendelian genetics—that competed with the standard Darwinian account for scientific consensus.[15] While Darwin's theory of evolution via natural selection never truly left the scientific mainstream, it was not until the rules of population genetics were given a formal mathematical basis in the 1930s and 1940s that Darwinism (or "neo-Darwinism," is it is sometimes called) emerged as the orthodox view in biology. It is worth bearing in mind, therefore, that while some of the biological theories discussed in this section may sound far-fetched to a modern reader, nearly all were considered well within the bounds of reasonable scientific discussion in their time.

The most salient non-Darwinian theory for our purposes was the widespread belief that evolution proceeds in a predetermined direction because of internal forces or innate tendencies acting on an evolutionary lineage. The broad label for this view is "orthogenesis," a term introduced and popularized in the 1890s by the German zoologists Wilhelm Haacke and Theodor Eimer, but it really describes a constellation of loosely similar approaches to evolution—many of which have much earlier roots—rather than a distinct school of thought. Late-nineteenth-century orthogenesis might, for example, invoke internal Lamarckian forces; or, on the contrary, it might explain directional evolution as the innate response of lineages to environmental pressures. One major feature of orthogenetic thought, though, was an emphasis on a cyclical view of evolutionary development; in this sense, orthogenesis was a return to Giambattista Brocchi's analogy between individual and species life cycles, which we examined in chapter 1.

Cyclical orthogenesis was especially popular among paleontologists, particularly in the United States and Germany, in part because of the perception that the fossil record did not display the smooth, even pattern of evolutionary transitions that Darwin predicted should be found there. Instead, what often emerged was a pattern of lengthy evolutionary stasis, in which little or no morphological change was observed in a lineage, followed by the abrupt appearance of new forms that might appear to be only distantly related to their putative evolutionary ancestors. This gave rise to speculation that natural selection alone might not be sufficient to explain the origin of genuinely new species or higher taxonomic groups, which many scientists—from the "Darwinian" T. H. Huxley to the "mutationist" Hugo de Vries to the geneticist Herman Muller—accounted for as evolutionary "saltations," or rapid jumps. While modern evolutionary theory has ruled out the possibility that large genetic saltations could produce viable new species, the notion that broad patterns of evolution observed in the fossil record fit a pattern of stasis and rapid evolution persists to this day, most prominently in Niles Eldredge and Stephen Jay Gould's hypothesis of "punctuated equilibria."[16]

The most common interpretation of extinction in an orthogenetic context was that the termination of evolutionary lineages represents a final, inevitable stage of decline in the life cycle of a species. Despite Darwin's apparent solution to the problem of extinction by treating it as nature's way of balancing the scales through natural selection, it remained a mysterious and contentious phenomenon to many observers. Darwin's explanation, for example, did little to illuminate why some long-lived and apparently well adapted groups disappear quite abruptly from the fossil record—trilobites, ammonites, and dinosaurs were favorite examples—nor why large ensembles of often heterogeneous taxa seem to have become extinct in coordinated fashion at certain points in the history of life.[17]

In the first case, the problem was less about proving that formerly successful taxa actually became extinct—the fossil record, notoriously incomplete as it may be, is nonetheless quite clear on this point—than it was about identifying rules, mechanisms, or even laws that could explain why one group survived while another did not. The answer of many orthogenetic interpretations was that *all* species have predetermined life cycles, and that it is possible to identify species in the final, senescent stage by observing certain characteristic trends in their morphology—such as gigantism, overspecialized anatomy, atrophied or vestigial organs, and so on. The second case reignited the debate about mass extinctions—episodes of widespread catastrophic extinction in a geological instant—that had been fairly quiet since Lyell's repudiation of Cuvier's theory of periodic revolutions. As paleontologists collected more fossils, it became increasingly clear that the major breaks in the history of life observed by Cuvier and others in the early nineteenth century (which became the basis for stratigraphic divisions in geology) were not going away, and that the possibility of catastrophic mass extinction, unpalatable as it was to Darwin, would have to be revisited. Broadly speaking, these were separate issues, and orthogenetic life cycles did not have much explanatory value for understanding mass extinction. Nonetheless, many authors did take these problems to be related, and together they formed the basis for a new geological view of extinction that emerged in the early twentieth century.

These issues were concisely summarized by the American paleontologist Alpheus Packard in an 1886 paper titled "Geological Extinction and Some of Its Apparent Causes," in which he remarked:

> The fact of extinction is indeed not less marvelous than that of evolution, and one cannot in these days feel satisfied that the solution of the problem lies in the theory of natural selection, which accounts for the preservation of species rather than their origin or extinction (Packard 1886, 29).

This essay was published in the journal *The American Naturalist*, which Packard himself had cofounded in 1867 along with Alpheus Hyatt and other naturalists sympathetic to neo-Lamarckian or orthogenetic evolutionary theories, and which was eventually purchased by the arch-Lamarckian vertebrate paleontologist Edward Drinker Cope. Packard's own interpretation of extinction was explicitly cyclical, and he proposed "a natural limit to the age of species as well as to individuals," noting that just as individual organisms experience "a youth, manhood and old age, so species and orders rise, culminate and decline" (Packard 1886, 40). He also described his views as "opposed to ultra-uniformitarian ideas," and while he was careful to stress that they had "nothing in common with the Cuverian catastrophic doctrine," he nonetheless observed that "the known facts of paleontology postulate long periods of quiet preparation, succeeded by more or less sudden crises, both local and general, to certain faunas or groups of animals, as well as individual species" (Packard 1886, 39). This statement may seem somewhat equivocal, and it was; the taint of Cuverian "catastrophism" was still powerful enough—as it would remain for decades—that even non-Darwinian evolutionary theories were wise to steer clear of it. But increasingly, paleontologists were open to acknowledging that relatively sudden events on a local, if not global, scale bore some responsibility for causing extinctions. In Packard's case, as in many similar views, such "sudden crises" could be invoked as the death blow that dispatched already senile species, rather than as the primary cause for their extinction. This conveniently also helped explain why, even in times of mass extinction, some groups were annihilated while others escaped unscathed.

While the fossil record for marine invertebrates such as mollusks and crustaceans was and continues to be the largest source of data for analysis of the history of life (because those organisms are more numerous and easily fossilizable than marine or terrestrial vertebrates), it is undeniable that large terrestrial animals—mammals, reptiles, birds, and of course dinosaurs—are the stars of paleontology. Public fascination with dinosaurs began when the group was first named in the mid-nineteenth century by the English comparative anatomist Richard Owen, and it has continued ever since. The late nineteenth and early twentieth centuries also saw a pronounced effort on behalf of museums and universities in Europe and especially the United States to fund expeditions in the hope of discovering new dinosaurs and collecting complete dinosaur skeletons that could be assembled for public display. This "great dinosaur rush" occupied considerable scientific as well as public attention, and it fueled a number of controversies, including the famous "bone wars" between the American paleontologists Cope and Othenio Charles Marsh.[18] It also helped elevate dinosaurs as the paradigm case for explaining extinction. The obvious question, in the late nineteenth century as today, was, Why did such a diverse and dominant group of animals perish in such an apparently short amount of time?

Absent some kind of Cuverian catastrophe, the most popular explanation was inevitable racial decline. This notion was present as far back as Owen's characterization of dinosaurs as slow, lumbering, cold-blooded reptiles who eventually had to give way to the smaller, nimble, and more intelligent mammals. In this context, dinosaurs easily fit a Victorian morality play about inevitable progress: being unable to keep up, they were simply left behind. This view began to change by the early twentieth century, however, as it became increasingly clear that mammals, generally small in body size and not terribly diversified at the late-Cretaceous termination of the dinosaurs' reign, hardly posed a competitive threat. As the great American vertebrate paleontologist Henry Fairfield Osborn put it in his authoritative *Age of Mammals* (1910), "There is little doubt that the extinction of the large terrestrial and aquatic reptiles, which survived to the very close of the Cretaceous, prepared the way for the evolution of the mammals" (Osborn 1910, 97). In other words, while the mammals benefited from the extinction of the dinosaurs, they could not possibly have caused it. Osborn himself was reluctant to assign a cause to the dinosaur mass extinction, although in this book and in writings on mammalian extinction he frequently discussed senescence as a possible cause of extinction.

Others, however, were less diffident about the subject. In his section report to the British Association for the Advancement of Science in 1909, the paleontologist Arthur Smith Woodward carefully noted that "the new race [mammals] did not immediately replace the old [dinosaurs], or exterminate it by unequal competition." But with equal confidence, he asserted that the dinosaurs were the victims of "racial old age," evidenced by "a superfluity of dead matter, which accumulates in the form of spines or bosses as soon as the race they represent has reached its prime and begins to be on the downgrade" (Woodward 1910, 464, 466). The prominent Yale University paleontologist Richard Swann Lull similarly argued in his influential textbook *Organic Evolution* (1917) that the dinosaurs suffered "racial death" because of extreme senility and overspecialization. In fact, he wrote, the dinosaurs had become so senescent at the time of their demise that "the marvel is, not that they died, but that they survived so long" (Lull 1917, 225). Likewise, Lull's Yale colleague Charles Schuchert, in his own popular 1924 textbook, stressed environmental changes as the probable source of extinction in a group "so highly specialized as were the Cretaceous dinosaurs"; but more broadly argued that since "just as individuals may in old age develop senescent characters, so frequently do the races. . . . When races are senile, or overspecialized, or are the giants of their stocks, they are apt to disappear with the great physiographic and climatic changes that periodically appear in the history of the earth" (Schuchert 1924, 497, 11–12).

The examples above merely exemplify a broad general opinion that racial senility was one of the leading causes of extinction of apparently well adapted groups in the history of life. The distinctive feature here is that, while during the Victorian era theories of extinction tended to reinforce a narrative in which nature steadily progressed through a competition in which "superior" forms replaced "inferior" ones, by the early twentieth century the broad picture looked less and less progressive. As the historian Peter Bowler notes, "The general feeling that the mammals got their chance to expand only when some external agency removed the dinosaurs suggests that the image of inevitable progress was now being heavily qualified" (Bowler 1996, 363–64).

This attitude was closely connected with contemporary ideas about human historical progress. In his four-volume *Outline of History* (1920), H. G. Wells devoted considerable attention to prehistory, and in particular to dinosaur extinction, which he described as "beyond all question, the most striking revolution in the whole history of the earth before the coming of mankind." Referring to the event as a "catastrophic alteration," Wells acknowledged that "as for the Mammals competing with and ousting the less fit reptiles . . . there is not a scrap of evidence of any such direct competition," and concluded that "first the reptiles in some inexplicable way perished, and then later on, after a very hard time for all life upon the earth . . . [mammals] developed and spread to fill the vacant world" (Wells 1920, 46–47). Many of the paleontologists we have already discussed also explicitly compared organic cycles of development to human ones, including an end phase of inevitable extinction. Packard wrote in 1887 that, as "species and orders rise, culminate and decline," so "nations have risen, reached a maximum of development and decayed." Meanwhile, Lull argued that the dinosaurs "do not represent a futile attempt on the part of nature to people the world with creatures of insignificant moment, but are comparable in majestic rise, slow culmination, and dramatic fall to the greatest nations of antiquity" (Packard 1886, 40; Lull 1917, 531–32). As Bowler again explains, "The growing debate over the causes of their [major groups'] decline and extinction . . . mark the revival of interest in a model of history that has strong parallels with the rise and fall of great empires in human civilization" (Bowler 1996, 436–37).

As we have seen, cyclical historical models were especially popular in Germany, and they reached an apotheosis in Spengler's *Decline of the West*, which was published right in the middle of these debates about organic extinction. It should be no surprise to learn, then, that cyclical theories of biological development—and of extinction resulting from inevitable racial decline—had special popularity among German-speaking paleontologists, and indeed remained popular well after they had begun to lose favor in Britain and the United States. The nineteenth-century embryologist Ernst Haeckel had already established a model of phases of evolutionary development in his 1866 *Generelle Morphologie der Organismen*, which he compared by direct analogy to the life stages of an individual organism: "We call the first stage of phylogeny, which is equivalent to the ontogenetic Anaplase, its time of blooming (Epacme), the second, which corresponds to the Metaplase, the flowering-time (Acme), and the third, which corresponds to the Cataplase, the wilting-time [*Verblühzeit*] (Paracme)." Epacme, acme, and paracme thus correspond to the birth, maturity, and senile stages of the life of an individual, and the last stage, paracme, which Haeckel explicitly associated with "old age" (*Greisenalter*) and "time of degeneration" (*Rückbildungszeit*), ultimately leads either to transmutation or to "total extinction" (Haeckel 1866, 321–22).

Haeckel's influence on the development of subsequent approaches to orthogenesis in Germany was quite significant, and a number of prominent early-twentieth-century German-speaking paleontologists expanded the cyclical notion into a broad interpretation of the history of life. Othenio Abel, an Austrian paleontologist active during the first several decades of the twentieth century, promoted an internal theory of evolutionary degeneration and extinction. Abel opposed a Darwinian, environmental interpretation of extinction, writing that "the degeneration of the species should be seen as a consequence of reaching the optimum of existence, and not as a consequence of particular changes in conditions of life"; and he argued that if it was not the only cause of extinction, degeneration was "certainly one of the most important" (Abel 1921, 59). Abel's conception of degeneration had an ideological component as well, and he was an early supporter of the Nazi party and a proponent of eugenic "race hygiene" arguments.

While Abel did not necessarily present a cyclical view of the history of life, his rejection of external influences on development in favor of internal forces or drives was emblematic of a distinctively German "*völkisch*" biological ideology that came to be associated with a tradition of German paleontology often referred to as "idealistic morphology." This is too large a subject to enter here, but the basic idea behind idealistic morphology—as developed by the poet and naturalist Johann Wolfgang von Goethe, and other Romantic-era biologists—was that variations in organic form are derived from a single, transcendental "blueprint" or archetype. In the later nineteenth century, however, the idea took on an explicitly evolutionary context, requiring the invocation of mysterious, vitalistic internal forces to explain the evolutionary development of organisms along pathways derived from a morphological ideal type.

The leading proponent of this new approach in the twentieth century was the German paleontologist Karl Beurlen, who, along with his contemporary Otto Schindewolf, was responsible for popularizing a cyclical, internalist theory of evolutionary development known as typostrophism, which dominated German paleontology for several decades. Typostrophism generally combined a version of idealistic morphology with a saltational view of species change (e.g., rapid production of new types) and, most important for our discussion, a cyclical view of evolutionary development. As Beurlen defined his approach:

> It is a general rule that the path of development within a related unit—
> and apart from whether it is a unit of higher or lower order—proceeds
> cyclically, in which the development from a beginning phase, with richer,

more variable, and more explosive morphogenesis passes into a phase of orthogenetic continuation, in which development is directional and predetermined and does not produce new types, to an endphase of overgrowth and degeneration of form, which therefore leads to extinction (Beurlen 1932, 76).

While Beurlen did not invent the term (Schindewolf would do so a few years later), this is the essence of typostrophism, which we can see is not terribly different from other orthogenetic theories we have considered.

What *was* perhaps somewhat different was the context. Writing in the early 1930s, Beurlen was deeply influenced both by Spenglerian cyclical views of human history, and by National Socialism. Beurlen was an avid member of the Nazi party, and a supporter of a movement that has been labeled an "Aryan biology," one tenet of which was that the biological environment and human society are analogous, each being held together by a complex web of interdependence that can be explained by basic "laws of life" (*Lebengesetze*) applying equally to both.[19] In this view, however, mechanical causality was to be rejected in favor of a more holistic understanding of relationships both in the development of human society and in nature, As Beurlen put it,

> It is not a simple causal relationship in which we can understand organic development; because the causality of the organic, which [the embryologist Hans] Driesch described with the term "wholeness" [*Ganzheit*] and Spengler with the term "fate" [*Schicksal*], is irreversible and characterized by the inevitable cycle of birth—youth—maturity—old age—death. The expression of this "causality of the individual" in phylogeny is the cyclical development process with its different phases (Beurlen 1932, 79).

In addition to invoking Spengler, Beurlen also drew on the Nietzschean concept of the "will to power," which he invoked as an explanation for the relationship between an organism and its environment. While he acknowledged Darwin's recognition of the essential role of struggle, he took exception with the notion that the struggle for life was merely a matter of a brutish "struggle for the feeding-trough or for expediencies and utilities," preferring a more ennobling "struggle for the power, which makes possible a characteristic self-differentiation independent of the environment" (Beurlen 1937, 223). Beurlen explicitly associated this biological will to power with ideological themes in National Socialism that emphasized the individual control and domination of the social environment. Spengler, himself no friend of the Nazis, had also drawn on a Nietzchian conception of "will" to explain the drive that brought civilizations to dominance. Despite ideological differences, then, Beurlen and Spengler shared a similar worldview: one that mixed elements of the Romantic idealism of Goethe and Haeckel, rejected mechanical notions of causality and especially Darwinism, and viewed historical development as a nonprogressive cyclical process in which collective entities, whether species or civilizations, passed through predetermined stages leading ultimately to senility and extinction.

Typostrophism found its fullest and most influential expression, however, not with Beurlen but with his colleague Otto Schindewolf, whose 1950 *Grundfragen der Paläontologie* (*Basic Questions in Paleontology*) was probably the most important work in German-speaking paleontology of the mid-twentieth century. Though they shared a commitment to an internalist, cyclical theory of evolutionary development, and had both established promising university careers during the 1920s and early 1930s, Beurlen and Schindewolf very much moved in opposite directions. Beurlen capitalized on his association with the Nazis to attain a leading place in German paleontology through prestigious appointments in the Reich Research Council and a professorship at the Ludwig-Maximillians-University in Munich. Schindewolf, on the other hand, refused to support National Socialism and was publicly attacked by Beurlen, losing his position at the Prussian Geological Survey in the process (he actually began writing the *Grundfrage* during the Second World War, but was unable to publish it until afterward). After the war, however, the situation changed dramatically: "denazification" stripped Beurlen of his positions and respectability, forcing him ultimately to emigrate to Brazil in order to continue his scientific career, while Schindewolf rose to become the leading paleontologist in Germany, first as a professor at the Humboldt University in Berlin, and later as professor and director of the Paleontological Institute at the University of Tübingen. Such was Schindewolf's status that, as late as 1970, Stephen Jay Gould recalled attending a lecture in Tübingen where a "hushed awe" surrounded his participation, and "not a single younger German paleontologist dared to question anything he said during the public forum" (Gould 1993, ix).[20]

Another difference between Beurlen and Schindewolf is that the latter, unlike many German paleontologists of his day, shied away from the overtly Romantic, even "mystical" tendencies in idealistic morphology, in favor of a more rigorous interest in empirical comparison of forms. As Schindewolf put it in the *Grundfrage* (which was first translated into English in 1993), while "the position we take here is morphologically idealistic inasmuch as it consciously sets up as the basis for its system only the morphological relationships among organisms," it was not the system "of Goethe and his pre-Darwinian successors, who saw in idealistic morphology the ultimate ideal of biological knowledge," nor did it regard morphology "as an end in itself and an ultimate goal for biology." Rather, Schindewolf regarded his morphological approach quite straightforwardly as

> purely empirical scientific research. It proceeds with rigorous objectivity from the real, natural data, from the existing forms and the graded, successive steps of their diversity, and arranges them according to logical principles in a graduated conceptual system (Schindewolf 1993, 410-11).

While this conceptual difference was significant, it did not prevent Schindewolf, at least in his works up through 1950, from endorsing a cyclical view of evolution strongly indebted to earlier orthogenetic theories and to the internalist historiography of Spengler and others (fig. 3.2). He introduced the term "typostrophism" in a 1945 paper where he laid out the three stages of the life cycle of a species. The first, "typogenesis," involved the rapid emergence of a new evolutionary type (and thus quite accurately led Schindewolf to be labeled a "saltationist"). The second, "typostasis," was an orthogenetic phase where the species developed, often for lengthy periods, in a progressive morpho-

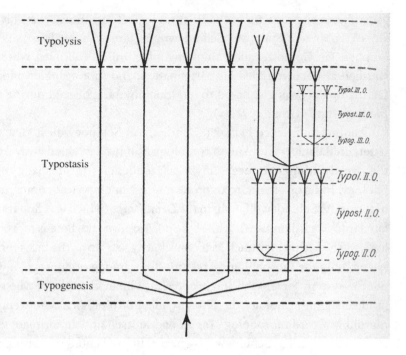

FIGURE 3.2 An illustration of the Typostrophic phases in Schindewolf's theory. The original caption states that during "the brief, final typolytic phase these subtypes lose their consistent morphological identity and produce all kinds of degenerative offshoots" before the lineage terminates through extinction. From Otto H. Schindewolf, *Basic Questions in Paleontology* (Chicago: University of Chicago Press, 1993), 202.

logical direction. And the final, "typolysis," was a period of senescence where the species degenerated to extinction. As Schindewolf explained in the *Grundfrage*,

> The third phase, *typolysis,* or the *dissolution of types* . . . brings each evolutionary cycle to a close. This phase is characterized by multiple indications of decline, degeneration, and the loosening of the morphological constraints embodied in the type. Overspecialization and gigantism in the lineages destined for extinction give this period its special mark (Schindewolf 1993, 193).

The phrase "destined for extinction" highlights the degree to which Schindewolf's views, as well as those of similarly-minded proponents of orthogenesis in this period, departed from the earlier Victorian reading of extinction as a mechanism for progressive evolutionary development. As Schindewolf quite bluntly put it, "This author believes that these phenomena [e.g., senescence] argue for a progressive aging of lineages and contradict the belief in unlimited progress held by Darwin and Lamarck" (Schindewolf 1993, 258).

The extent to which Schindewolf's attitude towards extinction of species reflected broader contemporary themes of cultural pessimism is difficult to pinpoint exactly. However, it is significant that, in a later essay from 1964 titled *"Erdgeschichte und Weltgeschichte"* ("Earth History and World History"), Schindewolf made an extensive argument for the analogy between natural and human history that drew directly on the cyclical model:

> Today it has been recognized, especially by O. Spengler and A. J. Toynbee, that the history of mankind does not so much run in a single track, as had been previously thought, but rather that a large number of original, independent cultures have existed, passing through their historical development in parallel, side-by-side or one after another, and sometimes without reciprocal interactions. All of these cultural bodies [*Kulturkörper*] have a limited lifetime and period of flourishing [*Blütezeit*]; they emerge, grow, fade and in each case are replaced by a new one.
>
> This is exactly what the history of life brings to mind for us. The flora and fauna also do not unfold in a linear, uniform historical course, but the development is realized independently and autonomously in numerous parallel phyla. Only the orders of magnitude are different. What in the history of life are phyla, classes, and orders correspond respectively in human culture cycles to races of a single human species and, if we add prehistory, to a few closely related species (Schindewolf 1964, 42).

In addition to Spengler, Schindewolf cited the British political historian A. J. Toynbee, whose twelve-volume *A Study of History* (1934–61) postulated a pattern of cyclical rise and fall of the world's great civilizations. Schindewolf went further than suggesting a superficial analogy, though, in proposing that the stages of civilization recognized by Spengler and Toynbee corresponded exactly with typostrophic counterparts in natural history: "A very remarkable parallel seems to me to exist, in that the cycles of human history occur in similar phases as we have become acquainted with from the development of geological and life history" (Schindewolf 1964, 44). Just as the first stage, typogenesis, brought about rapid, saltational organic change, as "O. Spengler, A. J. Toynbee, R. Coulborn and many others have shown, . . . civilizations arose respectively through a revolutionary act. These revolutions, that created the new type of culture, took place in a very short time." Likewise, Schindewolf argued that the final, typolitic stage had its parallel in human history:

> Various authors have often described in similar terms that originality would subside, the creative imagination and power would ebb away, no further possibilities for blossoming would be achieved, etc. The cultural body breaks down into smaller units that at the point of their cultural apex sink back down to primitive stages. These are the features of our typolytic phase. The initial indication of decline sometimes conceals itself under the mask of the seemingly greatest blossoming of power. That is, according to Toynbee, for example, the case with the mighty pyramids of the fourth Egyptian dynasty, which to the same extent can be placed on the same level as the monstrous dinosaurs. Through the outbreak of a new revolution, the dying culture may under certain circumstances recover and continue life through another cycle. Otherwise it expires or is supplanted by another culture, but in each case only after it had already internally eroded and collapsed, as is consistent with what took place with the displacement of reptiles by the mammals at the Cretaceous-Tertiary boundary (Shindewolf 1964, 24).

Conclusion

Interest in cyclical interpretations of the history of life persisted in Germany well after the Second World War, thanks largely to the influence of Schindewolf on the subsequent development of German paleontology. This influence has been described as inhibiting, since it effectively prevented German evolutionary theory from keeping in step with developments in Britain and the United States that, beginning in the 1940s, saw a fairly decisive shift back towards a Darwinian paradigm of adaptation and selection.[21] Indeed, writing a year before the publication of Schindewolf's *Grundfrage*, the American vertebrate paleontologist George Gaylord Simpson declared, "Races, or groups of organisms in general, do not seem to have any such life pattern. . . . Still less do they seem to have an inherent growth pattern or metabolic system which brings them to maturity at definite times and which dooms them to death from the internal ravages of old age" (Simpson 1948, 188). Simpson was speaking from his experience studying the evolution of mammals (especially horses), as well as from his perspective as one of the major framers of the Modern Evolutionary Synthesis. The Modern Synthesis—especially as articulated by two of Simpson's colleagues, the geneticist Theodosius Dobzhansky and the population biologist Ernst Mayr—had little tolerance for the Romantic conceptions of "force" or "will," or any of the other mysterious internal evolutionary mechanisms popular with orthogenesis and its fellow travelers.

At the same time, the issue of extinction would remain contentious for biologists and paleontologists for many more decades. Whether or not extinction was seen as the result of intrinsic life cycles or external selection pressures, the general sense—among paleontologists, at least—was that the Darwinian model of gradual competitive replacement was inadequate to explain certain phenomena in the fossil record. A major issue which had already cropped up in the pre-Synthesis years was the question of whether mass extinctions are a regular feature of the history of life. A number of paleontologists—including Osborn, Schindewolf, and Simpson—acknowledged that violent environmental events might play at least a local role in producing episodes of widespread extinction; and a very few—Schindewolf and Harry Marshall, for example—were open to the possibility that such events might be global and catastrophic, perhaps triggered by some extraterrestrial mechanism. After the Second World War a chorus of new voices would be added to these early speculations, and catastrophic mass extinction would finally reenter the mainstream of paleontological theory. In part this was because of new data and new interpretive frameworks; the advent of digital computers and multivariate statistical analysis allowed for new approaches to "reading" the fossil record. I will also argue, however, that broader cultural forces played a role in the more widespread acceptance of "catastrophism"—none greater than the splitting of the atom and the detonation of nuclear weapons over the cities of Hiroshima and Nagasaki. Living in a world where global nuclear annihilation was just the push of a button away produced a culture of anxiety that spilled into other areas as well: into fears of social collapse from population explosion, environmental catastrophe as the result of pollution, cataclysmic climate change produced by human industry, and the disintegration of traditional political structures and social mores through violent revolution. As Hobsbawm has put it, "Mass catastrophe, and increasingly the methods of barbarism, became an integral and expected part of the civilized world," teaching us "by the experience of our century to live in the expectation of apocalypse" (Hobbsbawm 1989, 330).

This chapter has suggested that clear roots of this later "catastrophic thinking" were planted in the decades before the Second World War. The first important shift was a reaction against the optimistic progressivism of the Victorian era, which we have followed in the literature, social and historical commentary, and science of the early twentieth century. The general pessimism toward progress that marked Modernist literature was also present in interpretations of human history, in theories of biological degeneration and decline, and in evolutionary thought. An additional feature highlighted in this chapter, "apocalypticism," had growing cultural currency, but did not translate directly to theories of biological extinction until after the Second World War. That shift will be discussed in the next chapter, where biological understanding of mass extinction will be placed in a broader cultural context of "postapocalyptic" thinking, which is distinguished from earlier forms of cultural pessimism in that the threat of potential catastrophe was no longer seen as metaphorical or avoidable, but rather understood to be inevitable and perhaps already underway. If in the early twentieth century apocalypse was a warning about a possible fate that might yet be averted, in the postwar, postapocalyptic context came the recognition that it might already be too late, and that we may be the agents of our own destruction. The science of extinction drew from this cultural context and informed it, both by demonstrating the historical traces of past catastrophes and by quantifying the consequences of the destructive path Western society had taken. What for Lawrence's readers was gloomy metaphor would become, for inhabitants of the later twentieth century, dismal fact: "The cataclysm has happened, we are among the ruins."

4

EXTINCTION IN THE SHADOW
OF THE BOMB

In the early hours of the morning on July 16, 1945, a small group of scientists and military observers witnessed something that the world had never seen before: a mushroom cloud blooming in the desert at the Trinity test site in the Jornada del Muerto desert in New Mexico. This was, of course, the first detonation of a nuclear weapon, and it was an event so unprecedented that some of the assembled scientists reportedly took morbid bets about whether it would set off a catastrophic chain reaction that would incinerate the entire atmosphere.[1] More soberly, the Manhattan Project scientific director J. Robert Oppenheimer later recalled his feelings at the time:

> We knew the world would not be the same. A few people laughed, a few people cried. Most people were silent. I remembered the line from the Hindu scripture, the Bhagavad-Gita; Vishnu is trying to persuade the Prince that he should do his duty, and to impress him, takes on his multi-armed form and says, "Now I am become Death, the destroyer of worlds." I suppose we all thought that, one way or another (Oppenheimer 1965).[2]

The Atomic Age had begun, and Oppenheimer was indeed correct: the world would forever be different in many ways. The specter of catastrophic annihilation that had shadowed the imagination of poets, scientists, historians, and politicians of previous decades had now become a reality.

As we have seen in the previous chapter, an apocalyptic sensibility was already well-established in the extinction imaginary of Western societies since the turn of the twentieth century. What was new about the atomic age was not the *idea* of a final civilization-ending catastrophe, but rather the *fact* of its imminence. This sentiment appeared time and again in contemporary commentary about life after the bomb. For example, in a widely read essay in the *Saturday Review* titled "Modern Man is Obsolete," the political journalist and peace advocate Norman Cousins wrote in 1945 that nuclear anxiety

> is a primitive fear, the fear of the unknown, the fear of forces man can neither channel nor comprehend. The fear is not new; in its classical form it is the fear of an irrational death. But overnight it has become intensified, magnified. It has burst out of the subconscious and into the conscious, filling the mind with primordial apprehensions (Cousins 1945, 5).

In a similar though more explicitly philosophical vein, the German theorist Karl Jaspers argued in his 1958 book *Die Atombombe und die Zukunft des Menschen* (translated in 1961 simply as *The Future of Mankind*):

> In the past there have been imaginative notions of the world's end. . . . But now we face the real possibility of such an end. The possible reality which we must henceforth reckon with — and reckon with, at the increasing pace of developments, in the near future — is no longer a fictitious end of the world (Jaspers 1958, 4).

This notion of the transfer of cataclysmic fear from the subconscious to the conscious, or from the fringes to the mainstream, was also highlighted by the prominent University of Chicago sociologist Edward Shils, who wrote in his 1956 *The Torment of Secrecy*:

> The atom bomb was a bridge over which the phantasies ordinarily confined to restricted sections of the population . . . entered the larger society which was facing an unprecedented threat to its continuance. The phantasies of apocalyptic visionaries now claimed the respectability of being a reasonable interpretation of the real situation (Shils 1956, 71).

In other words, as historian Spencer Weart puts it, "Nuclear weapons gave the twentieth century's nihilism a dismal solution. Immediately upon hearing the news from Hiroshima, sensitive thinkers had realized that doomsday—an idea that until then had seemed like a religious or science-fiction myth, something outside worldly time—would become as real a part of the possible future as tomorrow's breakfast" (Weart 1988, 392). The consequences of this social transformation—or perhaps collective psychological transformation in mass society—were profound and far-reaching. In politics it ushered in an age of "rational" paranoia symbolized in such cultural touchstones as "mutually assured destruction" and the "Doomsday Clock."[3] In mass media and literature it took the form of a heightened, almost resigned fatalism that has been described as a "postapocalyptic" mentality characterizing the works of authors as various as Walter Benjamin, J. G. Ballard, and Richard Matheson; and it was found equally in high-culture treatises and in popular entertainments.[4] In science, it opened the door for a reconsideration of the central topic of this book, extinction, as a potentially catastrophic threat of vital personal concern to every member of the human species. As Jaspers put it, the central threat imposed by the atomic age was "the extinction of life on the surface of the planet" (Jaspers 1961, 4).

This chapter will follow the tactic of the book as a whole so far by using political culture as a lens through which to understand the science of extinction, and vice versa. The most striking observation is that, beginning in the 1950s, the biological understanding of extinction underwent a slow but ultimately profound transformation that saw the gradual acceptance of a catastrophic model of mass extinction in paleontology and ecology as the best explanation of major changes in the diversity of life in the past—and perhaps in the future. This resulted in what would ultimately be described in the 1980s as the emergence of a "new catastrophism" that took hold in mainstream science, but it has clear origins in the culture and science of the decades immediately following the Second World War. As in the previous examples I have presented, this was not a straightforward matter of cause and effect; cultural anxieties did not "produce" a scientific catastrophism any more than new ideas about mass extinction generated social and political unease. Rather, the extinction imaginary of the 1950s and 1960s presents us with a tapestry in which a number of key themes are interwoven. These included, but were not limited to, the threat of sudden catastrophe (nuclear or otherwise), large-scale social unrest, increased awareness of environmental degradation, a discourse of cultural pessimism in the arts and humanities, the emergence of ecological theories that highlighted interconnectedness and fragility in ecosystems, and a scientific (and pseudoscientific) "catastrophism" around extinction.

The era after the Second World War has also retrospectively been labeled by many observers as "postmodern," a designation that has resonances with the topic of this chapter. While the term itself was first coined by Jean-François Lyotard in 1979, evidence of what Lyotard called the "postmodern condition" extends back to the immediate postwar period or even earlier.[5] Many of the central themes in postmodernity—a suspicion towards grand narratives, radical subjectivity, pronounced irony, a critique of late capitalism, and a pervasive discourse of disorientation—have roots in the literature and philosophy of Modernism, especially in the writings of Friedrich Nietzsche, Ludwig Wittgenstein, and Martin Heidegger.[6] Postmodernism can then be seen as an extension or outgrowth of the literary and philosophical Modernism and existentialism discussed in the previous chapter, with a couple of key qualifications. While Modernist authors frequently commented on the disintegration of traditional structures of meaning, many nonetheless harked back to the apparently firm certainties of an earlier age (evident in the romantic pastoralism found equally in poets such as Yeats and Eliot and in social commentaries by Brooks Adams and Ignatius Donnelly), or expressed hope for a revitalized civilization.[7] Postmodernism, in contrast, is characterized by a deeper sense of hopelessness or fatalism, as well as by an abandonment of earlier Western narratives of historical progress.

This is not the place to delve deeply into the topic of Postmodernism, but it is worth noting that many observers regard the horrific events of the Second World War—in particular, the Holocaust and the bombing of Hiroshima and Nagasaki—as watershed moments in the break between modernity and postmodernity. Events such as these were often referred to as "unthinkable," and the postmodern era therefore is seen as a period of time when formerly unthinkable events had become a reality. This sentiment resonates especially with some of the comments about the nuclear age presented above—where "fictions" or "phantasies" about the catastrophic end of humanity had become real. In religious terms, this corresponds to a transition from a premillennial theology, which anticipated a coming crisis as an opportunity for rebirth or renewal, to a postmillennial one, which regarded the world as already and irrevocably fallen.[8] In more secular language, we might speak of a distinction between apocalypticism and postapocalypticism, the latter of which the literary scholar Theresa Heffernan describes as a realization that we now "live in a time after the apocalypse, after the faith in a radically new world, of revelation, of unveiling" (Heffernan 2008, 6).

Postapocalyptic thinking took a very literal form in fictional imaginings, both literary and cinematic, of the aftermath of a nuclear war or environmental disaster—a genre that expanded dramatically from the 1950s onward. But these literal depictions of the aftermath of apocalypse had a strongly metaphorical element as well. For example, images of bombed cities and radioactive seas invoked contemporary realities of overcrowding and urban decay or industrial pollution that were increasingly becoming the focus of public and political concern. In other words, while the atomic bomb was a tangible symbol of impending catastrophe, it alluded to a broader culture of catastrophism and an extinction imaginary that emerged after the war and took many other forms. One consequence of this was that it opened cultural space for new ideas—many of which were progressive, such as the civil rights movement in the United States, the decolonization of European empires, and relaxed sexual and moral standards. But it also occasioned backlash and anxiety toward cultural change. This was true in science as well; the 1960s and 1970s, which have been described as a period of radical social change, also saw an antiestablishmentarianism among scientists, which manifested both in increased political activism by scientists and in a more permissive culture towards formerly heterodox ideas.

One such heterodox idea was mass extinction, which had been broadly rejected for nearly a century by mainstream paleontologists and geologists. While still by no means a widely accepted notion during the 1950s and 1960s—as the ruckus over Immanuel Velikovsky's "pseudoscientific" historical catastrophism, discussed below, will highlight—potentially catastrophic episodes of mass extinction in the earth's past became more frequent topics for discussion in paleontology and ecology. Although it would be too much to claim that the reconsideration of mass extinction was a direct product of nuclear fears, it is impossible to overlook the dramatic increase in the cultural currency of the term "extinction," which was frequently invoked as the catastrophic—and global—consequence of nuclear war. This had a circular reinforcing effect. On the one hand, nuclear annihilation provided a vivid image of the reality of world-altering physical cataclysm; on the other, empirical recognition of the reality of geological mass extinctions, which began to take hold in the late 1950s, gave historical validation to doomsday prophecies. And as time went on, models of the mechanisms and ecological consequences of catastrophic extinctions became the basis for predicting the effects of nuclear and ecological catastrophes of the present or future—though this will primarily be a topic for later chapters.

Finally, a new ecological understanding of the interconnectedness of life—and the rise of notions like the "ecosystem" and the "biosphere"—gave a more concrete conceptual vocabulary for describing the role of diversity in the natural world than had existed previously. In particular, ecologists began to theorize the relationship between ecological diversity and stability, arguing that diversity could be seen as a hedge against environmental or adaptive disruptions—and potential extinctions. This helped create a new, positive valuation of biological diversity in ecology and evolutionary biology, as well as a new sense of the fragility of the environment and the risks posed by unchecked human intervention. It also helped cement the notion that human beings are an intrinsic part of the global ecosystem and are subject to the same ecological forces that govern all other organisms—on whom we rely for the survival of our own species. This manifested itself in the consolidation of the modern environmental movement, and focused attention on crises involving industrial pollution and exponential population growth, which were clearly linked to both notions of ecological stability and the threat of mass extinction. It is in the scientific and political culture of the late

1960s that we see the clear roots of the late-twentieth- and twenty-first-century science and politics of biodiversity and extinction.

Nuclear Armageddon and the Future of Humankind

The rhetoric of potential nuclear catastrophe became a persistent feature of Western cultural discourse almost as soon as the announcement that the US B-29 bomber *Enola Gay* had dropped an atomic bomb on the Japanese city of Hiroshima on August 6, 1945. The world was not entirely unprepared for the event; ever since the publication in 1914 of H. G. Wells's novel *The World Set Free*, which imagined the consequences of a world war fought with primitive atomic weapons, fictional accounts of nuclear war or disaster had become a feature of speculative fiction and commentary. In 1940, Robert A. Heinlein published a short story titled "Blowups Happen" in the pulp magazine *Astounding Science-Fiction*, which described the tense atmosphere in a fictional nuclear power plant. The magazine's editor, John W. Campbell Jr., was fascinated by the theme of atomic disaster, and during the 1940s he encouraged his authors to explore the theme in their fiction. Campbell and his contributors drew inspiration from publicly available documentation of nuclear fission in scientific literature.[9] Fictional accounts of the time closely paralleled public warnings from scientists, and may have become something of a feedback loop; the Hungarian-born nuclear physicist Leo Szilard, one of the chief architects of the first nuclear reactor and a prominent Manhattan Project contributor, later admitted that he was inspired to pursue fission by the writings of Wells and others.[10]

Nonetheless, once the reality of nuclear weapons became public, anxiety about the possibility of sudden nuclear Armageddon spiked nearly instantaneously. As Weart comments, the "idea of apocalyptic power cropped up everywhere at once, like dormant seeds sprouting under a sudden rain" (Weart 1988, 104). In the summer of 1946, only a year after the bombings of Hiroshima and Nagasaki, the Committee on the Social Aspects of Atomic Energy of the US Social Science Research Council commissioned a national survey of Americans' attitudes about nuclear proliferation and war. The results of the survey bear out Weart's assessment: some 64 percent of respondents reported being concerned about the danger of an atomic attack on the United States, and 29 percent described the chances of themselves or a family member being killed by a nuclear weapon as either "very great" or "fairly great" (Cottrell and Eberhart 1969, 107–8). It is noteworthy that these surveys were conducted before the heyday of nuclear paranoia in the 1950s, associated with the infamous "duck and cover" civil defense drills.

From the very start, anxieties about nuclear disaster were a feature of the American popular psyche. These anxieties were no doubt stoked by grim commentaries in highly visible newspapers and magazines, which immediately cast the invention of nuclear weapons as an existential threat to humankind. For example, in a 1945 editorial in *Life* magazine, three prominent nuclear scientists described atomic forces as "responsible for the life and death of the stars" and warned that nuclear weapons are "a threat to the very existence of us all." Beyond this existential threat, they also pointed to the psychological impact of "a world in which atomic weapons will be owned by sovereign nations, and security against aggression will rest on fear of retaliation." They predicted that this would result in "a world of fear, of suspicion and almost inevitable final catastrophe" (Hill and Simpson 1945, 23–24). Norman Cousins, whose 1945 warning from the pages of his magazine *Saturday Review* is mentioned at the beginning of this chapter, argued that the advent of nuclear bombs heralded a new age in which the threat of "extinction" hung like "a blanket of obsolescence not only over the methods and the products of man but over man himself" (Cousins 1948, 5). This theme of epochal change in the Western mentality was echoed in other popular outlets as well. In a 1950 *New York Times* opinion piece titled "What the Atomic Age Has Done to Us," Michael Amrine contended that the bomb "underscores" a deeper lesson, that "civilizations can perish," and argued that it "attacks directly the belief almost unconsciously accepted by Western man: progress is inevitable. . . . The death of this idea is the most important death forecast by Hiroshima." Amrine suggested a "new humility in place of that pride which had been a concomitant of our belief that all evolution was upward. . . . The mushroom as a symbol radiates ideas more capable of chain reactions than neutrons" (Amrine 1950).

The connection between the atomic age and a new, pessimistic vision of the future for humankind was taken up in academic and philosophical discourse as well, featuring in the writings of European intellectuals including Hannah Arendt, Walter Benjamin, and Jacques Ellul. Perhaps the most explicit of such expressions was Karl Jaspers's *Future of Mankind*, which was widely read and reviewed on both sides of the Atlantic. Jaspers's main consideration was whether the threat of nuclear catastrophe would produce "a revolution in our way of thinking" to avert a potential crisis.[11] Throughout the book, Jaspers clearly identified the danger posed by nuclear weapons as "extinction"; again and again, he alluded to "the threat of total extinction," "the extermination of life," and "the destruction of mankind, of life itself" (Jaspers 1961, 6, 52, 2). While the chief aim of the book was to promote "a new politics" and "a call for reflection" that might avoid disaster, Jaspers was also quite stark in his assessment of the threat. "Now, mankind as a whole can be wiped out by men," he wrote. "It has not merely become possible for this to happen; on purely rational reflection it is probable that it will happen" (Jaspers 1961, 3). Ultimately, he concluded, humanity needed to recognize that a crucial historical turning point had been reached:

> In the past, the worst disasters could not kill mankind. . . . Life went on. Remnants led to new beginnings. Now, however, man can no longer afford disaster without the consequence of universal doom—an idea so novel, as a real probability, that we hesitate to think it through (Jaspers 1961, 318).

Not all analysis of the threat of nuclear war explicitly took the existential threat to humanity for granted, however. One of the iconic works of the nuclear age was Hermann Kahn's weighty treatise *On Thermonuclear War*, which was published in 1960. Kahn was a researcher at the RAND corporation—the chief US think tank for strategic analysis during the 1950s and 1960s—and he was tasked with a "quantitative analysis" of possible scenarios involving nuclear exchange. As he wrote in the preface, the book "examines the military side of what may be the major problem that faces civilization, comparing some of the alternatives that seem available and some of the implications in these choices" (Kahn 1960, preface). In more than six hundred chilling pages, Kahn outlined a cost-benefit analysis of a number of hypothetical scenarios, complete with a tally of the tens of millions of "megadeaths" that would result from the exchange. Nonetheless, he firmly believed that nuclear war would not necessarily mean the end of humanity, or even of democratic society. There would still be a society to rebuild, and he challenged the notion that, in the aftermath of nuclear conflict, "the survivors will envy the dead." Despite Kahn's calm assurances, the book was received with anything but relief. Most public attention was drawn to the later chapters in which Kahn contemplated a hypothetical "doomsday machine"—a device that could, at the push of a button, end all life on earth. Kahn intended this example to underline his thesis about the deterrence of so-called "mutually assured destruction," but in fact it had the opposite effect, capturing public anxieties about mad scientists that had long been the stuff of science fiction. This was used to dramatic effect in Stanley Kubrick's 1964 film *Dr. Strangelove; or, How I Learned to Stop Worrying and Love the Bomb*, in which the title character—an amalgam of several notable scientists and strategic analysts, including Kahn—presents the military with an actual doomsday device which is accidentally triggered at the end of the movie.

The cultural discourse of catastrophe, as presented especially in film and fiction, mirrored the new, more pessimistic tone of conversations around apocalypse after the Second World War. While fictional accounts of the end of the world had been circulated over the previous decades, as discussed in chapter 3, they burst from the fringes of culture and into the mainstream during the 1950s. This, again, presents something of a chicken-and-egg problem: While the growing popularity of apocalyptic and postapocalyptic science fiction undoubtedly reflected broader cultural and political anxieties of the time, changes in mass media also brought speculative stories to a much wider audience. For one thing, the film industry on both sides of the Atlantic changed in significant ways during and after the war. In both Britain and the United States, the film industry supported the war effort by producing propaganda and patriotic films, many of which depicted scenes of wartime destruction and casualties. Newsreel footage also brought the physical devastation of war home to American audiences who were geographically removed from scenes of actual conflict. Images of the London Blitz and the devastation of occupied Berlin were viewed by millions of theatergoers, and while the US government tightly controlled access to footage of the aftermath of Hiroshima and Nagasaki, newsreels did present images of the complete destruction of both cities, revealing postapocalyptic landscapes of utter devastation that were eerily absent of living people.

The 1950s also saw the rise of the television industry, with an estimated six million televisions in US homes by 1950, and sixty million by 1960.[12] While this made films even more accessible to many Americans, it also presented a challenge to the film industry, which responded by enticing moviegoers with more extravagant productions, provocative topics, and better special effects. Not surprisingly, many films of the 1950s and 1960s dealt, either directly or indirectly, with themes of war and catastrophe. Some movies attempted to depict the aftermath of nuclear war in literal and realistic terms. One of the earliest of these was the Columbia Pictures film *Five* (1951), which followed five survivors of a nuclear war struggling in a postapocalyptic landscape. While *Five* may have been the first such film, it was soon joined by others, such as 1955's *The Day the World Ended* (directed by a young Roger Corman, who later rose to prominence as the "king" of B-movies), 1959's *On the Beach* (adapted from the best-selling 1957 novel by Nevil Shute), and 1962's *Panic in Year Zero*. In 1964, Stanley Kubrick's *Dr. Strangelove* showed that nuclear annihilation could even be the subject of black comedy: one of the film's most memorable scenes depicts the cowboy-hatted Major Kong riding an atomic warhead as it descends towards its target, and the film concludes with a montage of nuclear explosions set to the sentimental Second World War song "We'll Meet Again."

But fictionalized accounts of nuclear war were only a small subset of the 1950s and 1960s films that dealt with themes of catastrophe and apocalypse. This era was a heyday of paranoid science fiction cinema, and films such as *When Worlds Collide* (1951), *The War of the Worlds* (1953), *World without End* (1956), *Invasion of the Body Snatchers* (1956),

The Time Machine (1960), *The Day of the Triffids* (1962), *The Last Man on Earth* (1964), *The Day the Earth Caught Fire* (1964), and many others depicted world-ending catastrophes in a variety of imaginative ways. As a genre, apocalyptic and postapocalyptic films have dealt metaphorically with a number of cultural and political themes, from ecological disaster and overpopulation (e.g., *Soylent Green*, 1973, and *Logan's Run*, 1976), to repressive totalitarianism (*Invasion of the Body Snatchers*, 1966, and *Fahrenheit 451*, 1966), to social unrest (George Romero's 1968 *Night of the Living Dead*, and sequels)—but they share a common undertone of pessimistic anxiety. Even when not being frightened by grim prognostications about actual nuclear proliferation, Cold War audiences apparently enjoyed being entertained by fictional portrayals of the collapse of civilization. At the very least, these films helped the public envision possible catastrophe in increasingly vivid detail.

The second major transformation in mass media culture was the rise in appeal of science fiction literature for mainstream audiences. During the 1930s and 1940s, science fiction was relegated largely to pulp magazines like *Astounding Science-Fiction*, which attracted a mostly juvenile male audience with stories of adventure on alien planets. Some literature of this era did deal with more mature themes, like global annihilation, as discussed above; but the circulation of these works was limited to a relatively small niche audience. In the 1950s, however, science fiction authors broke into the mainstream, as traditional book publishers began releasing speculative fiction in hardcover formats that opened new readerships in libraries and bookstores. The result was a flood of apocalyptic sci-fi literature onto popular consciousness. Some authors, like the former British military officer John Wyndham, were responsible for multiple entries. Wyndham became famous with his *The Day of the Triffids* (1951), a novel about a species of aggressive ambulatory plants that wipe out humankind, but he also penned the nuclear war novel *Tomorrow!* (1954) and the postapocalyptic survival tale *The Chrysalids* (1955). Likewise, the noted British dystopian author J. G. Ballard began his career with a string of novels imagining the end of civilization as the result, variously, of destructive winds (*The Wind from Nowhere*, 1961), climate change (*The Drowned World*, 1962, and *The Burning World*, 1964), and bizarre ecological disaster (*The Crystal World*, 1966).

Many of these novels focused on the experiences of survivors, either alone or in small groups, in frightening postapocalyptic landscapes. For example, Richard Matheson's *I Am Legend* (1954) depicted a lone survivor in a post-plague world of hungry vampires, and has been the basis for three film adaptations. Wyndham's *Day of the Triffids* also follows the attempts of a plucky group of survivors to seek refuge from murderous plants, while *The Chrysalids* imagines a society many centuries after a cataclysmic nuclear war. This latter theme was the subject of a number of other novels, including Walter M. Miller Jr.'s *A Canticle for Lebowitz* (1960), which follows a postapocalyptic society over thousands of years of rebuilding, and Pierre Boulle's *Planet of the Apes* (1963), in which astronauts journey to a distant planet where a humanlike race has been driven to primitive savagery and is enslaved by intelligent simians (the film version introduced time travel, and located the story on the earth of the future).

More so than films (although many of these novels were adapted to the screen), science fiction literature dealt with themes of alienation and despair, and often, unlike their film adaptations, ended on a pessimistic note. *A Canticle for Lebowitz*, for example, concludes with the suggestion that civilization is doomed—à la Oswald Spengler—to a cycle of destruction and rebirth, while *I Am Legend* ends with the death of the protagonist. Novels also allowed for extended authorial digressions or monologues that explored the causes and consequences of social decay and war, explicitly projecting modern anxieties onto fictionalized catastrophes. Take, for example, Wyndham's *Chrysalids*, where the long-ago disaster is referred to only as "the Tribulation," and the world's former inhabitants (i.e., we) are described as "only ingenious half-humans, little better than savages; all living shut off from one another." As one character explains to another:

> They could never have succeeded. If they had not brought down Tribulation which all but destroyed them; then they would have bred with the carelessness of animals until they had reduced themselves to poverty and misery, and ultimately to starvation and barbarism. One way or another they were foredoomed because they were an inadequate species (Wyndham 1955).

Novels such as these reflect not just a paranoia about nuclear proliferation or ecological disaster but also a deep disenchantment with earlier narratives of human progress and technological advancement. The message is that something is fundamentally wrong with humanity—"an inadequate species"—and that where some glimmer of hope for improvement is held out, it can only be achieved through a profound transformation of human society. When, upon seeing the remnants of the Statue of Liberty at the end of the film version of *Planet of the Apes* (1968), Charlton Heston's unlucky astronaut George Taylor cries out "You maniacs! You blew it up! Ah, damn you! God damn you all to hell!" he is echoing Jaspers's sentiment that "the end is either the extermination of life or the transformation of man and the human condition, so that physical conflict ceases" (Jaspers 1961, 52).

Earth in Upheaval: Catastrophism in Science and Pseudoscience

In the 1950s, fanciful stories of world-shattering catastrophes were not limited exclusively to science fiction. In 1950, the Russian-born psychoanalyst Immanuel Velikovsky created a sensation when he published the book *Worlds in Collision* with the respected trade and textbook publisher Macmillan and Company. While largely forgotten today (except in some corners of the Internet), Velikovsky's book was an immediate bestseller—as well as, in historian Michael Gordin's words, "one of the greatest publishing scandals of the postwar period" (Gordin 2012, 22). Velikovsky—who had no formal training in geology, astrophysics, archaeology, history, or any of the other topics considered in his book—was inspired by earlier authors such as Ignatius Donnelly to examine the mythology and scriptures of ancient civilizations for evidence that the earth has been subject to immense, worldwide catastrophes at various points in human history. His thesis, as stated at the beginning of *Worlds in Collision*, was "1) that there were physical upheavals of a global character in historical times; 2) that these catastrophes were caused by extraterrestrial agents; and 3) that these agents can be identified" (Velikovsky 1950, ix).

Specifically, Velikovsky claimed that around 1500 BCE a comet was ejected from Jupiter toward the earth, causing major disturbances in the earth's magnetic field and a reorientation of the earth's axis, unleashing meteor storms, tidal waves, and earthquakes before settling into orbit around the sun as the planet Venus. Among other events, this cosmic visitation was alleged to have been the source for the biblical passage in the Book of Daniel where Joshua commanded the sun to "stand still" (since the earth's rotation would have been temporarily interrupted). Velikovsky claimed to have found similar passages in contemporary mythologies of Greece, Egypt, Asia, and Mesoamerica. A précis of these ideas was presented by the journalist Eric Larrabee in a breathless article in *Harper's* magazine titled "The Day the Sun Stood Still" in January 1950, leading to several months of eager anticipation of the book's publication. While the article noted that many of Velikovsky's ideas were unorthodox, Larrabee nonetheless concluded that the work applies "all the apparatus of learning—from astronomy and physics to folklore, religion, classical literature, archaeology, geology, paleontology, biology, and psychology" to the "awesome task . . . of applying the techniques of scholarship and psychoanalysis to the entire human race" (Larrabee 1950, 26).

Most scientists, however, did not share Larrabee's enthusiasm. Even before the book was published, a furious campaign was launched against Macmillan in an effort to quash it, most prominently led by the Harvard astronomer Harlow Shapley. Immediately following the publication of Larrabee's *Harper's* article, Shapley wrote to Macmillan's editorial department to report that "a few scientists with whom I have talked to about this matter . . . are not a little astonished that the great Macmillan Company, famous for its scientific publications, would venture into the Black Arts without rather careful refereeing of the manuscript." He added that Velikovsky's thesis "is the most arrant nonsense of my experience," and expressed his "great relief" upon hearing rumors (unfounded, as it turned out) that Macmillan had canceled publication plans.[13] Macmillan editor James Putnam quickly wrote back to correct Shapley's misapprehension, and to assure him that "we are publishing this book not as a scientific publication, but as the presentation of a theory which, it seemed to us, should be brought to the attention of scholars in the various fields of science with which it deals." Putnam added that he expected there would "be a great diversity of reaction to the book" (Putnam to Shapley, January 24, 1950).

The entire correspondence between Shapley and Macmillan editors has been preserved publicly on a website devoted to Velikovsky's ideas (http://www.varchive.org/), and it makes entertaining reading.[14] Despite Shapley's ominous warnings about irreparable damage to Macmillan's scientific reputation, *Worlds in Collision* was indeed published in April 1950, and, as Putnam had predicted, reactions to the book were "diverse." Or, rather, they were split quite dramatically between those of scientists and highbrow intellectuals, who violently denounced the book, and the general book-buying public, who couldn't get enough of it. While sales figures are difficult to obtain, it is fairly certain that "millions" of copies were sold during Velikovsky's time in the sun between 1950 and the mid-1970s.[15] *Worlds in Collision* entered the *New York Times* bestseller list at number fourteen on April 16, 1950. By the next week it was number three, and by May 7 it was number one among nonfiction books—a ranking it held for nine weeks. Ultimately, it stayed on the *Times* list for thirty-one weeks during its initial run, and was continuously in print for decades, becoming—like similar works by Erich von Daniken (*Chariots of the Gods*) and L. Ron Hubbard (*Dianetics*)—popular among college-age and countercultural audiences in the 1960s and 1970s.

Unsurprisingly, the scientific community took a much dimmer view of Velikovsky's book. Reviewers in fields from astronomy to geology to classical archaeology were "unanimously negative" in their assessments, and focused on both major flaws (e.g., violations of the laws of gravity) and minor ones (misreadings of ancient texts).[16] To put it bluntly, Velikovsky's argument is completely implausible. Were an object the size of Venus to pass anywhere remotely near the earth, the consequences would be far more violent than Velikovsky proposed, and individual effects such as a temporary suspension of the earth's rotation are, from a physical point of view, impossible. Velikovsky, however, refused to back down, and in 1955 he published a follow-up, *Earth in Upheaval*, which expanded his argument to a broad geological theory of catastrophism.

Needless to say, this book was no more successful in convincing scientists, but it did maintain his popular momentum.

Leaving aside the scientific reaction for now, it is noteworthy that even Velikovsky's vehement critics recognized the cultural appeal of *Worlds in Collision*. For example, in his review in the *New Yorker* magazine, Alfred Kazin described the book as "extraordinarily unconvincing," and "preposterous and intellectually primitive to an extreme," and he lamented the general state of education among a public who would eagerly embrace such nonsense. But he also noted that as a "pathetic, ominous, and superstitious piece of work by a man whose thinking is completely dominated by cataclysms, catastrophes, and global disturbances," the book "fits only too well into the intellectual melodrama of this period" (Kazin 1950, 103). The real reason for the book's appeal, Kazin reasoned, was not in its scientific claims, but rather "that man is always on the brink of universal destruction, and that the most he can be is a recording agent of these prodigious disasters" (Kazin 1950, 104). Indeed, Kazin continued, Velikovsky's argument played "right into the small talk about universal destruction that is all around us now," and encouraged a passivity and pessimism in the face of incomprehensible forces. "These days," Kazin concluded, "even as we sit on the brink and wonder if all of us yet may go over, we can always read our fate in advance" (Kazin 1950, 104–5). Velikovsky himself seemed to realize and encourage such connections between his discussion of ancient catastrophes and the modern climate of geopolitical crisis. In the introduction to *Worlds in Collision*, he reflected:

> The years when *Ages of Chaos* [a separate book detailing textual evidence for catastrophes] and *Worlds in Collision* were written were years of a world catastrophe created by man—of war that was fought on land, on sea, and in the air. During that time man learned how to take apart a few of the bricks of which the universe is built—the atoms of uranium. If one day he should solve the problem of fission and fusion of the atoms of which the crust of the earth or its water and air are composed, he may perchance, by initiating a chain reaction, take this planet out of the struggle for survival among the members of the celestial spheres (Velikovsky 1950, ix).

While *Worlds in Collision* generally drew negative responses from astronomers and physicists, who were outraged at the book's preposterous claims about celestial mechanics, it did not engage directly with any central issues in contemporary mainstream science. There were no debates in the 1950s (or, for that matter, in the 1850s or the 1750s) about the gravitational effects of planets passing close by one another. Geology, though, was a different matter. Toward the end of *Worlds in Collision*, Velikovsky noted that "in the present volume geological and paleontological material was discussed only occasionally," but he promised to take up those topics more thoroughly in a future work.[17] In his next major book, *Earth in Upheaval*, Velikovsky expanded his theory of cosmic catastrophe to a broader geological catastrophism, and extended the cycle of upheaval back into deep prehistory. He also found a new publisher; after the furor over *Worlds in Collision*, Macmillan had decided to cancel their publishing agreement with Velikovsky, leading him to turn to Doubleday, which was more than happy to have his business.

In *Earth in Upheaval*, Velikovsky presented a theory of cyclical mass extinctions reminiscent of Georges Cuvier's cycles of "revolutions." In fact, while he criticized Cuvier's vague explanations for the mechanisms responsible for his revolutions, Velikovsky positioned himself very much as the heir to Cuvier and other nineteenth-century catastrophists. In explaining the long dominance of the "uniformitarian" view in geology, he pointed explicitly at cultural factors: "No wonder in that climate of reaction to the eruptions of revolution and the Napoleonic Wars the theory of uniformity became popular and soon dominant in the natural sciences" (Velikovsky 1955, 21). Now, however, he believed that there was sufficient evidence to overturn that paradigm once and for all, and to demonstrate that "the extermination of great numbers of animals of every species, and of many species in their entirety, was the effect of recurrent global catastrophes" (Velikovsky 1955, 210). In making his case, Velikovsky presented very little evidence that would not have been available to a nineteenth-century geologist; his chief witnesses were geological features like "erratic" boulders (deposits left behind by retreating ice sheets), geological unconformities (tilted sequences of strata), continental upthrust, climate change, ice ages, and the like. He drew on virtually no contemporary paleontological studies, and indeed seemed to have a positive disdain for recent research. Despite two decades of paleontological research into extinction, in the foreword to the 1977 edition of *Earth in Upheaval* Velikovsky claimed he saw no reason to alter the 1955 edition's text. Toward the end of the book, Velikovsky nonetheless asserted:

> The fact that the geological record shows a sudden emergence of many new forms at the beginning of each geological age does not require the artificial explanation that the records are always defective; the geological records truly reflect the changes in the animal and plant worlds from one period of geological time to the next. Many of the new species evolved in the wake of a global catastrophe, at the beginning of a new age, were entombed in a subsequent paroxysm of nature at the end of that age (Velikovsky 1955, 233).

Intentionally or otherwise, here Velikovsky was treading closer to an area of genuine scientific debate: whether the geological record should be viewed as an "imperfect document," as Darwin had urged his readers, or rather as a mostly complete text whose pages could be read literally. In the *Origin of Species*, Darwin had famously proclaimed, "We have no right to expect to find in our geological formations, an infinite number of those fine transitional forms, which on my theory assuredly have connected all the past and present species of the same group into one long and branching chain of life" (Darwin, 1859, 301). The reason for this, he argued, was that the fossil record is imperfect: "I look at the natural geological record as a history of the world imperfectly kept, and written in a changing dialect" (Darwin 1859, 310). While most paleontologists after Darwin accepted this dim view of their data source, a vocal minority had persisted in the belief that discontinuities in the fossil record—moments where major groups either disappeared suddenly or emerged in a geological instant—were a valid biological "signal," and not the artifact of a poor record.[18] By the late 1940s, the paleontologist George Gaylord Simpson had joined his influential voice to those efforts, arguing in *Tempo and Mode in Evolution* (1944) that "the face of the fossil record really does suggest normal discontinuity at all levels," and that even apparent "incompleteness is an essential datum and . . . can be studied with profit" (Simpson 1944, 99, 105).

Simpson did not focus significant attention on the problem of mass extinctions, but his protégée and close colleague at the American Museum of Natural History (AMNH), a young invertebrate paleontologist named Norman Newell, took up the topic enthusiastically. Newell, who would have a long career at the AMNH during which he would make foundational contributions to evolutionary paleobiology, marine paleoecology, and statistical analysis of the fossil record, is best remembered for championing the reality of mass extinctions at a time when they were still viewed with suspicion, if not outright hostility, by the majority of the paleontological profession.[19] He also embraced Simpson's view that the fossil record is an adequate source of data for broad evolutionary conclusions. Writing in 1952 on the subject of "periodicity in invertebrate evolution," he noted that while the record "is neither complete, nor uniformly good, . . . the record of fossil invertebrates is an impressive one, and probably is an adequate sample of the evolutionary history of the better known groups" (Newell 1952, 371–72). Newell's comments here mark an important turning point in the history of paleontological study of extinction: while previously much attention had been given to the spectacular disappearances of "charismatic" vertebrate groups such as the dinosaurs—and, to a lesser extent, more recent extinctions of large mammals like the mastodon—from this point on the problem of extinction would center on marine invertebrates like trilobites and mollusks. This is not to say that scientists or the public lost their fascination with dinosaur extinction—far from it, as we will see in the next chapter—but rather that as a source of data, marine invertebrate fossils, which have been preserved in quantities many orders of magnitude greater than vertebrate remains, offer a much better statistical sample on which to base theoretical conclusions.

In his 1952 paper Newell took for granted that the invertebrate fossil record revealed "mass extinctions of marine genera on a global scale," but he did not probe the causes or consequences of these events.[20] This changed in 1956, when he published a paper titled "Catastrophism and the Fossil Record" in the journal *Evolution* (notably, he chose a journal widely read by biologists rather than a more narrowly specialized paleontological journal). This essay was framed as a response to several papers on mass extinction by Otto Schindewolf that hypothesized global mass extinctions caused by bursts of cosmic radiation. While Newell did not accept Schindewolf's explanation, he did agree that "enigmatic, apparently world-wide, major interruptions in the fossil record. . . . are real, approximately synchronous, and are recognizable at many places in different parts of the world," and that "critical events in the history of life evidently were responsible for these world-wide revolutionary changes" (Newell 1956, 97). Newell hesitated, though, to label these mass extinctions as "catastrophic," since he felt that they could be explained as the cumulative effects of more gradual environmental trends, such as sea level changes. Nor was he comfortable with the "hypothetical cosmic agencies" proposed by Schindewolf, which Newell felt violated the "time-tested scientific procedure to avoid, if there is a practical alternative, hypothetical solutions, no matter how tempting, that depend on highly speculative and untested premises" (Newell 1956, 100).

Newell's response to Schindewolf—and indeed his approach to the problem of mass extinction throughout his career—demonstrates an important characteristic: Newell was, on the whole, an extremely careful and even conservative scientist who avoided speculation at all costs, and who repeatedly subjected his own findings to rigorous second-guessing and statistical testing. This is one reason why, in an era of popular excitement about speculative theories like Velikovsky's, Newell was relatively immune to being stuck with the much-feared label "crackpot." Newell was a widely respected scientist with impeccable credentials and an institutional affiliation that shielded him from suspicion of ulterior motives, which helped him, virtually singlehandedly, to establish the respectability of scientific investigation of mass extinction. Both his credentials and his measured approach, then, distinguished Newell from other writers about extinction at the time. For example, when the American spongiologist M. W. de Laubenfels, notably not a paleontologist, published a 1956 paper in *Journal of Paleontology* that posited an asteroid impact as the source of the extinction of the dinosaurs, he received virtually no response.[21] Despite the apparent reasonableness of de Laubenfels's arguments—he pointed out physical evidence of major impacts in the earth's past, along with astronomers' estimates of the relative frequency of near-earth asteroid encounters—his hypothesis was treated merely as idle speculation. Geologists and paleontologists are conservative by nature, and it would take overwhelmingly dramatic evidence for a similar hypothesis to be generally accepted some twenty-five years later (as we will see in the next chapter).

This characterization is accurate, at least, for the Anglo-American scientific community, where speculative catastrophic theories were relegated to popular books like Velikovsky's, or to only slightly more respectable works like geochemist Allan Kelly and astronomer Frank Dachille's 1953 book *Target Earth*, or Belgian mathematician and amateur geologist René Gallant's similar *Bombarded Earth* (1964).[22] In the Soviet Union, however, several paleontologists, including N. S. Shatskij, V. I. Krasovskiy, and I. S. Shklovskiy, explored possible extraterrestrial extinction mechanisms, though their work was never translated and therefore failed to make an impact on the wider profession.[23] The main standard-bearer for catastrophic extinction remained Schindewolf, who despite being viewed with suspicion by many American scientists was still the most influential paleontologist in Germany. His final major publication on the subject was the 1963 paper "Neokatastrophismus?," which was cited in a number of Newell's later publications. But even Schindewolf's endorsement of "catastrophism" was somewhat equivocal, as signaled by the question mark in the title; he accepted the term "only so long as it is made clear that the ideas it portrays have hardly anything to do with Cuvier's catastrophism," and broadly argued that since cosmic radiation is merely the mechanism for inducing mutations that accelerate episodes of racial senescence in some groups, "this is not conceived in the terms of a natural catastrophe that has betaken the whole of the Earth with great suddenness and absolute simultaneity" (Schindewolf 1977, 10, 14).[24]

Nonetheless, despite his cautious nature, by the early 1960s Newell was prepared to be more aggressive in his claims. In 1962, as outgoing president of the Paleontological Society, he had the opportunity to present a major address titled "Paleontological Gaps and Geochronology." Here he focused mainly on the question of whether major breaks in stratigraphic sequence—Darwin's "missing pages" from the history of the earth—were real or artifacts of poor preservation, and he argued that at least two of the most celebrated such breaks—at the end of the Permian some 250 million years ago, and again 65 million years ago at the end of the Cretaceous—were likely correlated with major mass extinctions. The next year, in 1963, he presented a major study of mass extinctions, "Revolutions in the History of Life," at a special symposium organized by the Geological Society of America, the proceedings of which were not published until 1967. Here Newell made the bold claim that "the purpose of this essay is to demonstrate that the history of life . . . has been episodic rather than uniform, and to show that modern paleontology must incorporate certain aspects of both catastrophism and uniformitarianism while rejecting others" (Newell 1967, 64). While he avoided endorsing traditional Cuverian catastrophism explicitly, Newell nonetheless emphasized the unpredictable nature of the history of life, and took aim at some basic uniformitarian assumptions, writing that "catastrophism rightly emphasized the episodic character of geologic history, the rapidity of some changes, and the difficulty of drawing exact analogies between past and present" (Newell 1967, 65).

In particular, this paper emphasized that mass extinction played a much more important role in evolution than had been normally credited. Noting that periods of mass extinction tended to be followed by "episodes of exceptional radiation" (i.e., bursts of accelerated evolution), Newell argued that mass extinctions were key events that cleared ecological space for new adaptive opportunities and evolutionary experiments. To illustrate this correlation, he published a graph in which major adaptive radiations were superimposed against mass extinctions (fig. 4.1).

This highlights another important feature of Newell's extinction studies, which was their quantitative methodology: Newell's conclusions about mass extinctions were based on an extensive evaluation of data on thousands of taxonomic groups (this paper focused at the taxonomic level of the family), which he analyzed statistically in order to determine relatively precise calculations for extinction rates at particular times. Mass extinctions were identified—and would continue to be in future extinction studies—as episodes where quantitative extinction

Percentage of Total Families

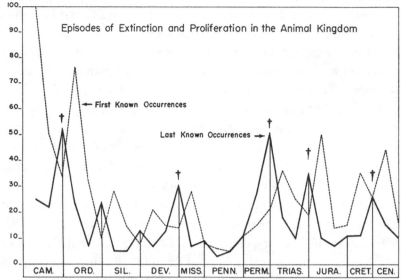

FIGURE 4.1 Graph of appearance (dotted line) and disappearance (solid line) of marine animals in the fossil record. The solid-line peaks represent the major mass extinctions at the ends of the Cambrian, Devonian, Permian, Triassic, and Cretaceous periods. Note that they are generally followed a short time later by episodes of increased diversification. Norman D. Newell, "Revolutions in the History of Life," in *Uniformity and Simplicity*, Geological Society of America special paper 89 (Boulder, CO: Geological Society of America, 1967), 79. Used with permission of the Geological Society of America.

rates were identifiably higher than in "normal" times. In other words, mass extinctions came to be redefined as statistical anomalies in taxonomic data, or as episodes when the standing diversity of life dropped below normal thresholds.

Another crucial feature of Newell's approach to extinction was that he explicitly understood mass extinctions as anomalous fluctuations in diversity. Traditionally, "extinction" has been a rather nebulous concept in biology. Its broad definition, as given, for example, in *Keywords in Evolutionary Biology*, is "a terminal event in the history of a population, species, or higher taxon" (Damuth 1992, 106). Right away, the confusion is evident: extinction would seem to result when the death of the very last member of some group occurs. The natural level at which this definition would make most sense would be the species, since species are often invested with some kind of concrete existence—they are "onto-logically real" or "natural kinds," as a philosopher would say. The death of an entire population, then, would seem only to be extinction if that population was the last of a particular species anywhere on earth. The "extinction" of higher categories (like genus or family) would also appear problematic, since those groups are generally considered less "real" or distinct than species. What we are saying when we say that a genus has become extinct is really that all of its constituent species have died out.

Furthermore, how do we detect extinction? Since extinction is defined as the absence of some entity, we face the old dictum that "absence of evidence is not evidence of absence." It can be very difficult to definitively establish truly that no living member of a species exists anywhere in the world, and indeed we often see news reports that a member of some previously extinct species has been "found," usually in some remote locale (famous examples, like the coelacanth, abound). From a paleontological perspective the task would seem more straightforward, since for more than two centuries paleontologists have kept detailed taxonomic records on when individual species and higher taxa appear in and disappear from the fossil record. But of course the fossil record is notoriously incomplete; and even if it is regarded as being reliable, the fact that a species drops out of a stratigraphic sequence—which may cover tens of millions of years—gives us very little information about when or why it died out. In addition, paleontology faces the problem of so-called "pseudoextinction": when a particular lineage ends because it has evolved into a new species (hence, we sometimes refer to birds as "living dinosaurs"). This underscores the fact that species are *not* really stable entities, but rather are taxonomic units composed of populations that share similar but not necessarily identical genetic traits, and which can be widely distributed geographically. "Speciation" occurs when biologists or paleontologists determine that members of a population have sufficiently differentiated either genetically or morphologically, respectively, from their peers. Needless to say, this is not always easy to pinpoint.

What this shows us is that from a biological or paleontological perspective, it is very difficult to precisely identify the circumstances or causes of extinction, which perhaps explains scientists' reluctance to study the phenomenon in great detail before the 1960s and 1970s. If it is difficult verging on impossible to precisely document the extinction of a particular species, it makes much more sense to adopt a statistical or probabilistic model of extinction: if species or groups of species are prevalent in the fossil record at some point, and at some later point are entirely absent, extinction most likely has taken place somewhere in between. The concept of *mass* extinction identifies cases in which some large number of groups (usually measured at the genus or family level, but ostensibly made up of individual species) disappear from the record in coordinated fashion, usually within a few million years of one another. The evidentiary threshold is somewhat lower for mass extinctions, because even if we are wrong about the timing of a few individual extinctions, we can still establish a statistical likelihood that many of them are real.

Mass extinctions are, therefore, more easily identifiable than individual extinctions, because they stand out against the "background" more clearly. And how is this detected? As has already been mentioned, since the early nineteenth century, paleontologists and geologists have continuously added to a census of life on earth over time, taking careful note of when species appear and disappear from the record. In the first instance, this information was used to document the diversity of life and to understand how life has evolved and how ecosystems function. Indeed, Newell's own extinction studies began, as discussed in his 1952 paper on periodicity in evolution, as an attempt to quantitatively assess diversity. He identified mass extinctions because he recognized that, while on the whole diversity has increased over the history of life, there are anomalous periods where diversity drops sharply. Mass extinction, then, came to be defined not just as a phenomenon most easily identified by statistical analysis of fossil data, but as an anomalous fluctuation in data about *diversity*. A "mass extinction" is a period of unusually low levels of standing diversity—and this is a concept that Newell helped popularize that is still relied upon today.

But merely identifying periods of sharp diversity loss doesn't tell us very much about how or why mass extinctions have taken place—

so Newell's studies opened up more questions than they answered. For most of his career, Newell argued, as he did in his 1963 symposium paper, that worldwide decline in sea level was the most likely culprit. As he put it, "It seems clear that rapid emergence of the continents would result in catastrophic changes in both terrestrial and marine habitats and such changes might well trigger mass extinctions among the most fragile species" (Newell 1967, 88). Note, however, that while he used the term "catastrophic" to describe the changes, he was not describing a single short-term event like an asteroid impact, but rather a process that might well have played out over tens of millions of years. Furthermore, since he regarded these extinctions to mostly affect "the most fragile species," he remained convinced, like Darwin, that extinctions had a strongly selective component. Newell's insights were the inspiration for many further studies of mass extinction, and the causes he assigned are still often widely regarded as valid, but they provided relatively little insight into cases where enormously widespread and broadly adapted groups, like the dinosaurs, disappear from the record in a geological in- stant. Still less did they settle the lingering, and at the time resurgent, debate between catastrophism and uniformitarianism.

In his 1963 symposium paper, Newell was quite careful to avoid step- ping into this larger debate or adding fuel to any wider cultural associa- tions with extinction. A term like "catastrophism is a term with an emo- tional connotation that implies calamity and destruction," he argued, "and as such it is not appropriate in any scientific context" (Newell 1967, 66). However, in 1963 he also adapted his extinction research to an essay in the popular magazine *Scientific American*, titled "Crises in the History of Life," in which he was notably less restrained. In par- ticular, Newell drew comparisons between historical mass extinctions and current environmental depredation and even the threat of nuclear war: "We are now witnessing the disastrous effects on organic nature of the explosive spread of the human species and the concurrent de- velopment of an efficient technology of destruction" (Newell 1963, 83). Newell pointed to a number of factors as contributing to rapid loss of species, including hunting, the destruction of habitats, pollution and in- secticides, urban sprawl, and invasive species, concluding overall:

This cursory glance at recent extinctions indicates that excessive predation, destruction of habitat and invasion of established communities by
man and his domestic animals have been the primary causes of extinctions within historical time. The resulting disturbances of community
equilibrium and shock waves of readjustment have produced ecological
explosions with far reaching effects (Newell 1963, 86).

This statement raises several important points. In the first place, it
marks the beginning of a trend—which continues to this day—in which
paleontologists, normally confined to matters of the deep past, used paleontological expertise about extinction and biodiversity to communi-
cate with the public about modern diversity crises. From this point on,
paleontologists would have an important voice in political and cultural
discussions about extinction and endangerment. Second, in his popular
article Newell more explicitly introduced an ecological logic for linking mass extinction with diversity. In describing mass extinctions of
the geological past, he explained that "the interdependence of living
organisms, involving complex chains of food supply, may provide an
important key to the understanding of how relatively small changes in
the environment could have triggered mass extinctions" (Newell 1963,
77). In other words, a small environmental change could have a snow-
ball effect, since the removal of one even apparently trivial or humble
component of the system could initiate a domino-like propagation
of ecological failure. As Newell put it, "No organism is stronger than
the weakest link in its ecological chain"—a lesson as potentially vital
for human survival as it was for that of the trilobites or the dinosaurs
(Newell 1963, 85). Furthermore, the local consequences of ecological
disruption could propagate in time as well: loss of diversity was always
followed by eventual ecological recovery, but what was lost in ecological or genetic diversity could never be truly recovered. An underlying
conclusion, therefore, was quite simple: "Extinction is an evolutionary
as well as an ecological problem" (Newell 1963, 86).

Diversity, Stability, and Extinction

What we see, then, in the development of scientific ideas about extinction by the mid-1960s is twofold: first, a recognition that mass extinctions may have been a recurrent feature of the history of life on earth, and second, awareness that a key to understanding the causes of past extinctions—and potentially predicting the consequences of future ones—lies in understanding ecological interdependence. The concept of diversity is important in both cases. Mass extinctions are defined as major depletions of standing biological diversity, and are measured by studying the history of organic diversity in the fossil record. But diversity is also important because in an ecological sense it can contribute to the stability or instability of a system. Remember that Darwin assumed that diversity had essentially remained constant over the history of life—there were no mass extinctions, but also no real threat of ecological collapse, since nature tended to replace species with organisms equally well suited for their particular environments.

In the early 1960s, the notion that diversity was an important component in the stability of complex systems was fairly new. Newell was not alone among paleontologists in invoking this explanation: for example, the paleontologist James Beerbower's widely used college textbook *Search for the Past* (1960) argued that species exist "in a rather delicate ecological adjustment to one another," and that "if someone upsets the applecart by being extinct—due, say, to climactic change—the whole system is likely to be unfavorably affected," potentially resulting in mass extinctions.[25] But this notion was part of a broader transformation—in ecology and genetics, particularly—that had taken place over the previous few decades, and it would have cultural as well as scientific ramifications.

At the heart of the matter are deep-seated biological assumptions about equilibrium and stability—the "balance of nature." As we saw in the first two chapters, nineteenth-century biologists tended to assume that nature remains in perpetual balance because of some kind of inherent regulating principle. In crude terms, this meant that nature tended to ensure that all available resources were maximized by placing organisms in "stations" appropriate to the needs and habits of each. In the earlier part of the nineteenth century, there was often the strong suspicion that this balance was divinely inspired; after Darwin, the principle of natural selection naturalized such notions by naming selection as the "agent" responsible for maintaining this balance. By Darwin's logic (followed by that of others, including Alfred Russell Wallace), natural selection ensured that the earth was always populated by a relatively stable diversity of species, since the zero-sum principle of competitive replacement meant that species were fighting for a finite number of environmental resources.

Much of this thinking was focused and crystallized with the emergence of the modern discipline of ecology, which was a late-nineteenth-century development.[26] The introduction of the niche concept—first coined by Joseph Grinnell in 1913, and later codified by Charles Elton in the 1930s and 1940s—created a language for talking about the relationship between organisms and their environments that allowed more precise, quantitative investigations. In its initial development, the niche concept hewed very close to Darwin's logic: according to Elton, the principle of "competitive exclusion"—the notion that only one species could occupy a particular niche in a local ecosystem—was a central feature in the balancing of ecological systems. This did not, however, mean that nature was static: on the contrary, Elton believed that competition meant that ecosystems were in constant flux, shuffling the species that occupied particular niches. As he put it in his classic textbook *Animal Ecology and Evolution*, "'The balance of nature' does not exist, and perhaps has never existed. The numbers of wild animals are constantly varying to a greater or less extent, and the variations are usually irregular in period and always irregular in amplitude'" (Elton 1930, 17). Importantly, Elton was strongly drawn to the idea that ecosystems tended towards *equilibrium*, which he opposed to classic notions of the balance of nature.

Two of the central concepts in ecology, then, as the discipline moved into the mid-twentieth century, were competition and equilibrium. The constant variation in numbers of particular organisms was seen as a function of competition—between individuals of a species, between species (predator-prey relationships, competition for scarce resources), and between all organisms and their environments—but the overall effect was that an ecosystem as a whole tended towards stability, in the sense that ecological niches were full and the food web was complete. As the Soviet ecologist Vladimir I. Vernadsky—one of the chief popularizers of the term "biosphere"—put it in 1944, "The single living organism recedes from view; the sum of all organisms, i.e., living matter, is what is important" (Vernadsky 1944, 487). This stability, though, is a tenuous arrangement: individual organisms and species are constantly under stress, and entire ecosystems can be threatened by environmental perturbations that can produce violent disruptions. As the historian Joel Hagen has observed, the tension around equilibrium in the ecological thought of the time reflected broader cultural perceptions: "The industrial society that Elton saw in nature, though basically stable, was at times subjected to unpredictable and violent disturbances," since "like human industrial societies, animal communities were not completely free from violent and unpredictable events" (Hagen 1992, 56–59).

Ecology needed a model for understanding this relationship between volatility and stability, and an important lesson was provided by contemporary genetics. During the 1930s and 1940s, the field of genetics—especially as directed toward questions of selection and evolution—developed a focus on the population as the key unit of study. One of the most important contributors to this shift was the Russian-born geneticist Theodosius Dobzhansky, who came to the United States in 1927 to work in the world-renowned research group led by Thomas Hunt Morgan, one of the founding figures in modern genetics, whose "Fly Room" at Columbia University was the source of multiple breakthroughs and Nobel Prizes in the unlocking of the mechanisms of heredity.[27] By the 1930s Dobzhansky had, following Morgan's study of fruit flies (*Drosophila melanogaster*), established his own reputation, and in 1937 he published a book titled *Genetics and the Origin of Species*, which is widely regarded as being one of the formative texts in the so-called Modern Evolutionary Synthesis.

Dobzhansky had focused most of his empirical research on examining how mutations moved through populations of flies, which led him to ponder the role of genetic heterogeneity—or diversity—in stable biological populations. This was one of the major topics of *Genetics and the Origin of Species*; in fact, the first chapter was titled, simply, "Organic Diversity." While Dobzhansky began by observing, "For centuries man has been interested in the diversity of living beings," he distinguished between a descriptive approach—"recording as accurately as possible the multitudinous structures and functions of the beings now living and of those preserved as fossils"—and a "nomothetic" (law-producing) one—that is, "an analysis of causes underlying the diversity" (Dobzhansky 1937, 3–6). The former approach, he argued, is the province of natural history, while the latter should be the goal of genetics. A central question, then, was why do populations of organisms—say, *Drosophila* flies in the wild—contain so much latent genetic variability (by which he meant chromosomal variations or latent mutations), while being physiologically (i.e., phenotypically) so similar?

As Morgan's group had determined, most significant *Drosophila* mutations proved to be harmful—offering no selective advantage at best, and being fatal at worst—yet the *Drosophila* genotype was filled with such latent variations. Might this inherent variability be seen, Dobzhansky mused, as "a destructive process, a sort of deterioration of the genotype that threatens the very existence of the species and can finally lead only to its extinction"? (Dobzhansky 1937, 126). This was certainly the fear of the late-nineteenth- and early-twentieth-century promoters of "racial degeneration" we examined in the last chapter, whose anxieties about a disastrous accumulation of negative hereditary traits in human populations helped instigate the eugenics movement in the United States, Britain, and Europe. However, Dobzhansky dismissed this as the perspective of "eugenical Jeremiahs," arguing precisely the reverse: as he put it, far from being a threat to the survival of a population, "the accumulation of germinal changes in the population genotypes is, in the long run, a necessity if the species is to preserve its evolutionary plasticity" (Dobzhansky 1937, 126). By "evolutionary plasticity," Dobzhansky meant the ability for a population to try out new phenotypic solutions to evolutionary challenges, which could be accomplished only if there existed, latent in the population's genotype, sufficient options (in the form of recessive mutations, for example) for new experiments. "The environment is in a state of constant flux," he wrote, "and its changes, whether slow or catastrophic, make the genotypes of the past generations no longer fit for survival. . . . Hence the necessity for the species to possess at all times a store of concealed, potential, variability" (Dobzhansky 1937, 126–27). In other words, genetic diversity, far from being harmful, was a storehouse of potential variation that could preserve a population or species should drastic environmental or adaptive changes take place. Or, as Dobzhansky grimly warned, "A species perfectly adapted to its environment may be completely destroyed by a change in the latter if no hereditary variability is available in the hour of need" (Dobzhansky 1937, 127).

As well as being a sharp rebuke to the ambitions of eugenicists — whose utopian visions usually imagined a homogeneous human society composed of only those with the "best" traits — Dobzhansky reasoning opened up a new line of argument for the value of diversity in many contexts. Dobzhansky himself was a lifelong advocate for racial equality, and often used similar arguments to undermine beliefs that superficial phenotypic differences among humans (like skin color) betokened quantifiable differences in behavior, intelligence, or the like. In fact, he stressed that human racial diversity was a positive attribute of the species, writing in his popular book *Mankind Evolving* (1962) of his "hope that mankind may eventually profit by this diversity more than it might have gained by monotonous sameness, even of the most 'advanced' kind" (Dobzhansky 1962, 286). But Dobzhansky's insight about the relationship between diversity — meaning adaptive flexibility — and stability had much farther-reaching influence. In a sense, this is one of the later twentieth century's most important (if often unexamined) cultural notions: that any complex collection of biological entities — whether a genetic population, an ecological system, or a human society — is made stronger and more resilient to change by having a "storehouse" of variability. Diversity, in other words, became reconceived as an inherent property of healthy collectives, and therefore came to hold inherent positive value.

This conception is evident in the kind of ecological thinking expressed by paleontologists like Newell and Beerbower when they wrote of "the interdependence of living organisms" or the "rather delicate ecological adjustment" of species as factors determining extinction or survival during mass extinctions of the geological past. It also became an explicit tenet of contemporary ecology, which during the 1950s and 1960s taking was its own "nomothetic" turn. One of the main figures in this era was the Yale ecologist G. Evelyn Hutchinson, who moved from an early career studying lake ecology following the tradition of Elton and others to an interest in an approach to a causal understanding of ecological relationships centered around abstract mathematical models.[28] Hutchinson established a thriving school of "theoretical ecology" around him at Yale, and his most prominent student, the mathematical ecologist Robert MacArthur, became a pioneer in an approach to ecology using simple mathematical models as heuristic devices for understanding complex ecological dynamics.[29] In a 1955 paper titled "Fluctuations of Animal Populations," MacArthur argued that a key component of ecological stability is a constant level of species abundance, which he expressed as the number of paths energy can take through a food web. Stability increases, he argued, as the number of links in the food web increase—in other words, as the number of distinct interrelated niches in an ecosystem are filled—because more of the energy in the system is being reabsorbed.[30]

Taking his student's reasoning a step further, in 1959 Hutchinson published a paper with the unusual title "Homage to Santa Rosalia" (whom he nominated as the patron saint of evolutionary biologists) and the subtitle "Why Are There So Many Kinds of Animals?" Noting that humans have been fascinated with the great diversity in the organic world for centuries, Hutchinson proposed an ecological answer, which he acknowledged was derived from MacArthur's paper: "There is a great diversity of organisms because communities of many diversified organisms are better able to persist than communities of less diversified organisms" (Hutchinson 1959, 150). He justified this claim on the grounds of MacArthur's reasoning about ecological thermodynamics, as well as for evolutionary reasons closely analogous to Dobzhansky's arguments, writing that "it is probable that a group containing more diversified species will be able to seize new evolutionary opportunities [more] than an undiversified group" (Hutchinson 1959, 155). Hutchinson's (and MacArthur's) thinking on this topic would prove to be enormously influential, especially in later paleontological analysis of diversification and extinction. It would—as we will see in the next chapter—provide a line of argument for explaining why certain groups appear to have been differentially affected by mass extinctions of the past. It would also give insight into the consequences of major extinction events, which by definition deplete, perhaps permanently, the global ecosystem's store-house of both ecological and genetic diversity. This last point would become central in debates about the consequences of the accumulating biodiversity crisis, but Hutchinson anticipated those later arguments by several decades. As he concluded in his 1959 paper, human activity has reduced global ecological diversity "in an indiscriminate manner," and we may only "hope for a limited reversal of this process when man becomes aware of the value of diversity no less in an economic than in an esthetic and scientific sense" (Hutchinson 1959, 156).

Indeed, by the early 1960s this ecological message had already begun to make its way to a wider public audience. The theme of "interconnection" was, as historian Thomas Robertson has argued, a common Cold War trope in politics as well as in science. Noting that President Eisenhower referred to "the basic law of interdependence" in his 1953 inaugural address, Robertson shows that discussions of interconnection were especially prominent in the rise of "neo-Malthusian" thinking, or concern about the ability of the earth to sustain an exponentially growing human population.[31] One of the most prominent members of this school was the Stanford University biologist and environmental activist Paul Ehrlich. Ehrlich is best known for his 1968 bestselling book *The Population Bomb* (cowritten, as many of his popular books would be, with his wife, Anne Ehrlich), which revived Thomas Malthus's grim late-eighteenth-century calculations about the relationship between population growth and the availability of food resources. Having received his PhD in 1957, Ehrlich was deeply steeped in both the populational approach of the Modern Evolutionary Synthesis and the new ecological thinking promoted by Elton, Eugene Odum, Hutchinson, and others; and he applied these lessons to his analysis of the problems he saw facing global society in the twentieth century.

While Ehrlich was aggressive in applying ideas from population biology and ecology to contemporary social problems, he was skeptical about what he saw as outdated ideas about the stability of nature. Indeed, in a 1967 paper coauthored with the Australian geneticist Louis

Charles Birch, he noted that while the notion of "a 'balance of nature' is commonly held by biologists," it is "difficult to explain why it persists in the writings of ecologists" (Ehrlich and Birch 1967, 97). The concept, Ehrlich and Birch argued, is a holdover of the fallacy that one-for-one replacement is a feature of ecological systems, and a misapplication of the notion of equilibrium. Yes, they conceded, ecosystems do tend to oscillate around a mean value, achieving a kind of equilibrium, but that does not mean they are stable; this equilibrium fluctuates in response to changing environmental factors, and major disturbances in any of the variables that constitute an ecological system can easily throw that system into disarray. This is most evident in the case of population size: while for decades ecologists had pointed to internal constraints on population growth—so-called density-dependent factors, which were thought to limit population size by reducing populations through attrition as resources were used up—Ehrlich and Birch argued that this would not necessarily prevent population explosions with disastrous ecological consequences. As Ehrlich and Birch bluntly concluded in their paper, "The notion that nature is in some sort of 'balance' with respect to population size, or that populations in general show relatively little fluctuation, is false" (Ehrlich and Birch 1967, 106).

This reasoning was the basis for *The Population Bomb*, which appeared the following year (fig. 4.2). One of the defining features of the book was its pessimistic tone: right from the start, Ehrlich warned that "the battle to feed humanity is over," and predicted that during the 1970s "hundreds of millions of people will starve to death in spite of any crash programs embarked upon now" (Ehrlich 1968). This distinguishes Ehrlich's book from earlier warnings about population explosion, such as Fairfield Osborn's 1948 *Our Plundered Planet*, which had generally adopted the view that a catastrophe could still be averted. In this sense, Ehrlich's views of population fit comfortably into what I have described as the postapocalyptic Cold War worldview, which regarded crisis as a foregone conclusion rather than one of several possible future outcomes. While Ehrlich did not predict that human extinction was a necessary result of global overpopulation, he nonetheless raised it as a possibility, arguing, "The birth rate must be brought into balance with the death rate or mankind will breed itself into oblivion" (Ehrlich 1968, pro-

FIGURE 4.2 The cover of Paul Ehrlich's best seller *The Population Bomb* (New York: Ballantine, 1971).

logue). Many of Ehrlich's claims have been challenged over the years, and in the early twenty-first century it is clear that his most dire prognostications did not come to pass. But it is undeniable that his views found a receptive public: since its publication *The Population Bomb* has sold more than two million copies, and it helped launch Ehrlich's career as an influential public intellectual.

One of the central features of Ehrlich's population arguments was a focus on the dynamics of ecological systems. "One of the basic facts of population biology," he wrote, "is that the simpler an ecosystem is, the more unstable it is" (Ehrlich 1968, 49). He argued that humankind had been systematically simplifying global ecosystems through industrialized farming and agriculture, the use of pesticides, and urban sprawl, and that this activity would have dire consequences. While he did not explicitly use the term, Ehrlich was developing what would become a lifelong concern with preserving biological diversity; and, as we will see in later chapters, he became a major voice in the biodiversity movement of the 1990s and beyond. Ehrlich also would be a public figure in scientific commentary about the dangers of nuclear proliferation, which he connected to population explosion in painting a dismal potential scenario in *The Population Bomb*. Projecting a political crisis that could result from the worldwide famines he predicted would take place during the 1970s, Ehrlich imagined the outbreak of nuclear war in the 1980s as the final capstone to the global crisis, leading to an extinction event from which "the most intelligent creatures ultimately surviving . . . are the cockroaches" (Ehrlich 1968, 78).

As we have seen in the case of nuclear anxieties earlier in this chapter, it is difficult to make causal claims about the relationship between scientific ideas and social fears during this period. Clearly, authors such as Ehrlich tapped into complex currents of popular anxiety at the time. Ehrlich's choice of the term "bomb" in the title of his book was deliberate, and was explicitly intended to invoke technological horror and fear of sudden catastrophe—which, in addition to nuclear holocaust, included mistrust of rapid social change and of new environmental movements.[32] Furthermore, projections similar to Ehrlich's had long been circulating in strategic analysis circles in Washington and elsewhere. As the environmental historian Jacob Hamblin has shown, modeling of catastrophic scenarios by RAND and other policy analysts was not limited to nuclear war; a whole host of environmental catastrophes—as the result of global climate change, crop failure, overpopulation, and other factors—had been researched as part of the broader Cold War program of risk assessment in the United States since the 1950s.[33] It was even suggested that some of these outcomes might be deliberately triggered as part of military strategy. As Hamblin argues, these projections influenced a counter-response by environmentalists, and were a key factor in the birth of the modern environmental movement.

One of the signature early works of that emerging movement was Rachel Carson's 1962 best seller *Silent Spring*, which alerted a mass public audience to the dangers of industrial pesticides, and which was published during the height of the Cuban Missile Crisis. The theme of mass extinction was a feature of Carson's writing, as was the new ecological perspective on the stability of nature. As Carson explained, the dangers posed by pesticides was not merely their toxicity to humans, but also their effect on the diversity of entire ecosystems, of which humans were active members:

> The balance of nature is not a *status quo*; it is fluid, ever shifting, in a constant state of adjustment. Man, too, is part of this balance. Sometimes the balance is in his favor; sometimes—and all too often through his own activities—it is shifted to his disadvantage (Carson 1962, 246).

Carson also emphasized that environmental conservation was not merely a matter of protecting the most visible endangered species; given the close interrelationships between all organisms, even tiny changes to our environment could have dramatic consequences. Here she was strongly influenced by Elton, who had devoted the final chapter in his 1958 book *The Ecology of Invasions* to the topic of "the conservation of variety." Taking an argument similar to those of MacArthur and Hutchinson, Elton warned that in "the exploited lands of the world we see a decrease in richness and variety of species" due to the use of pesticides and industrial monoculture, and he stressed his belief "that conservation should mean the keeping or putting in the landscape of the greatest possible ecological variety—in the world, in every continent or island, and so far as practicable in every district" (Elton 1958, 154–55). In *Silent Spring*, Carson quoted the botanist LeRoy Stegman, who had observed, "A few false moves on the part of man may result in the destruction of soil productivity and the arthropods may well take over" (Stegman 1960).[34] Ultimately, Carson argued, the specter facing humankind was a potential mass extinction of its own devising: "Along with the possibility of the extinction of mankind by nuclear war, the central problem of our age has therefore become the contamination of man's total environment with such substances [pesticides] of incredible potential for harm" (Carson 1962, 8).

Conclusion

The picture painted in this chapter of the two decades following the Second World War is undoubtedly one of pervasive gloom and crisis. Mass extinction, I have argued, was a central conduit between scientific and popular anxieties about a world that seemed to many to be perched on the brink of catastrophe. The specter of nuclear annihilation was the obvious touchstone in this culture of anxiety, and it both contributed to and was reinforced by biologists' and paleontologists' investigations of the dynamics of extinction and biological diversity. This extinction imaginary, I have also argued, can broadly be described, in distinction to the one that characterized the first half of the twentieth century, as postapocalyptic, meaning that the threat of ultimate disaster and extinction was viewed no longer as a potential future outcome, but rather as something already underway. Part of this sense came from the reality of events like the Holocaust and the bombings of Hiroshima and Nagasaki, but on the scientific side it also came from a growing acceptance of the fact that, perhaps many times in the geological past, great catastrophes had been visited on the earth, with perhaps permanent ecological and evolutionary consequences.

At the same time, the message was not entirely pessimistic. The development of ecology—and its influence on a growing popular environmental consciousness—contributed to a greater sense of interconnectedness between peoples of all nations, between organisms in ecosystems, and between human beings and their fellow inhabitants of the earth. This theme of interconnection and interdependence helped create a new valuation of diversity as an essential component in the stability of complex systems. Humans had not yet entirely given up on the hope for a better future, but it was becoming clear that the way forward would involve much greater humility and responsibility toward one another and our environments. This was an essential precondition for the eventual recognition of a biodiversity crisis and a "sixth mass extinction," and for the beginnings of a new culture of catastrophe—one might call it "post-postapocalypticism"—that was less anthropocentric and less focused on the single devastating event than previous cultures had been. I will discuss this in the sixth and final chapter of this book, where I will argue that, following the breakup of the Eastern bloc and the end of the Cold War, twenty-first-century societies have adopted a longer-term perspective in their catastrophic thinking. The threats facing humanity, global climate change and biodiversity depletion, have largely replaced sudden nuclear annihilation as the dystopian outcome, and the sense of humans' relationship to the rest of the organic world has adopted the rhythms of the deeper scales of geological time. This is reflected in the current widespread belief that we are now living in the era of the "Anthropocene."

But first we need to follow our story through the 1970s and 1980s, and in particular we must trace the further development of scientific understanding of mass extinction—which reached its apex in the early 1980s—and the escalating fear of violent nuclear confrontation and its aftermath on the popular imagination. These topics will be the subject of the next chapter, and will set the stage for a consideration of how we have come to understand extinction and diversity to be so vitally linked today.

5

THE ASTEROID AND THE DINOSAUR

On November 20, 1983, some hundred million Americans sat down to watch the ABC Sunday Night Movie *The Day After*. The film, which was billed as an "authentic" depiction of the effects of nuclear war, captured public attention and sparked a national (and international; it was eventually screened in theaters abroad) discussion about the consequences of nuclear exchange in a way that no single event had done since the dropping of the atomic bombs on Hiroshima and Nagasaki. It was arguably one of the most significant television events of the twentieth century: its Nielson rating (46.0), market share (62 percent), and number of households (more than 38 million) still rank the initial showing as one of the most watched broadcasts of all time (surpassed only by a handful of Super Bowls, the final episode of *M*A*S*H*, and the reveal to the "Who Shot JR?" cliffhanger of *Dallas*).[1] It had a particularly profound effect on the millions of school-aged children who gathered with their families to watch it: one of the film's most indelible moments, played out in total silence, in which a mushroom cloud erupts from a spectacular flash of light, bathing stranded motorists fleeing Kansas City in an eerie orange light, became a regular feature of nightmares for a generation.

To viewers today, many elements of the film might come off as hokey—even as they did at the time—and the special effects were largely unimpressive. The cast of characters was standard TV movie fare, including an assortment of stereotypical "average Americans" (recently engaged high school sweethearts, an earnest college student, a noble doctor, a stoic farmer) coping with the realization of imminent attack and then the devastating aftermath. The film featured few recognizable stars (Jason Robards, as the heroic doctor, being the notable exception), and the production was beset by problems that included battles between the director, Nicholas Meyer, and network executives and censors over the movie's length and pacing and the graphic nature of some scenes. But despite all of this, *The Day After* was and remains a powerful and haunting film, and it crystallizes a defining moment in the cultural history of nuclear anxiety. From this point on, it became increasingly difficult to avoid acknowledging what some political observers and scientists had been warning for a number of years: that a large-scale nuclear exchange was not something that modern civilization could survive in any meaningful sense, and that it might well bring about the extinction of the human species.

While the most pessimistic outcomes were only hinted at in the film itself (which limited its dramatic scope to the inhabitants of Kansas City and Lawrence, Kansas), viewers were made all too aware of the potential consequences of total war. Even before the final credits rolled (over a scene in which Robards collapses in a silent embrace with a fellow survivor amid catastrophic destruction), a text appeared onscreen announcing, "The catastrophic events you have just witnessed are, in all likelihood, less severe than the destruction that would actually occur in the event of a full nuclear strike against the United States." The broadcast was immediately followed by a live ninety-minute ABC News debate moderated by Ted Koppel that featured cold warriors including Henry Kissinger, William F. Buckley Jr., and Robert McNamara, along with the holocaust survivor and author Elie Wiesel and the astronomer Carl Sagan, who had gained great public visibility through his 1980 PBS series *Cosmos*. Many millions of viewers stayed tuned for an exchange that was, though at times contentious, surprisingly univocal in its condemnation of nuclear proliferation. While some panelists derided the film itself (Buckley described it as "antinuclear propaganda"), virtually all agreed that it highlighted the need for sane, bilateral reductions in nuclear stockpiles.

One of the most memorable moments of the post-viewing debate came when Sagan used a vivid metaphor to characterize Soviet-US strategic deterrence, describing the superpowers as "two implacable enemies" sitting in "a room awash with gasoline," each holding thousands of matches. But it was another comment, in response to a provocation by Buckley, that stands out with more significance in retrospect. Turning to the cameras, Sagan sternly intoned his "unhappy duty to point out that the reality is much worse than what's been portrayed in this movie," describing a bleak scenario:

> The nuclear winter that will follow even a small nuclear war . . . involves a pall of dust and smoke which would reduce the temperatures—not just in the northern mid-latitudes, but pretty much globally—to subfreezing temperatures for months. In addition, it's dark, the radiation from radioactivity is much more than what we've been told before, agriculture will be wiped out, and it's very clear that beyond the one or two billion people who would be killed directly . . . the overall consequences would be much more dire, and the biologists who've been studying this think that there's a real possibility of the extinction of the human species from such a war (Sagan 1983a).

The nuclear winter scenario Sagan was describing would become central to ongoing debates about arms reduction and deterrence. The idea itself had been unveiled by Sagan and colleagues in simultaneously published popular (*Parade* magazine) and scientific outlets (*Science*) less than a month before *The Day After* was broadcast, and the hypothesis was immediately connected with the political message of the film.[2] As multiple editorials pointed out, the nuclear winter scenario not only portended a potential global extinction event, but could also be triggered by a smaller exchange (e.g., one hundred megatons) than full-blown nuclear war—thus altering the strategic calculus assumed in most assessments of "limited exchange."[3]

The threat of widespread global extinction conjured by Sagan and others coincided with another spectacular and spectacularly popular scientific development in the early 1980s: the discovery, by a team led by the father-son duo Luis and Walter Alvarez, of an anomalous layer of iridium at the stratigraphic break between the Cretaceous and Tertiary periods of geological history.[4] The Alvarez hypothesis, as it came to be known, was both a scientific and popular bombshell. The initial study, published in 1980, suggested that the only mechanism for depositing large quantities of iridium (a rare element on the upper layers of the earth's crust) in the sediment would have been an impact with a large, extraterrestrial body—an asteroid or comet perhaps ten kilometers in diameter. Furthermore, the timing of this impact appears to have been quite significant: the Cretaceous-Tertiary (K-T) boundary dates to roughly sixty-five million years ago, precisely when the dinosaurs—not to mention a number of other terrestrial and marine groups—abruptly disappear from the fossil record.[5] While there was some precedent for hypotheses involving spectacular extraterrestrial triggers for mass extinction, as we have seen in previous chapters, no prior theory had the solid empirical grounding and testability of the Alvarez hypothesis. The effect was to catalyze discussions of catastrophic mass extinction that had been building momentum towards the end of the 1970s, and to galvanize opinions among geologists, paleontologists, and astrophysicists about the nature and consequences of extinction events that would radically change understandings of the history of life.

The Alvarez hypothesis also coincided dramatically with late–Cold War nuclear anxieties, projecting the fate of the dinosaurs as an object lesson for humanity in countless popular articles and opinion pieces. This, however, was more than mere coincidence. Consider the descrip-

---

FIGURE 5.1A–D A series of paintings from 1983 depicting an artist's conception of the asteroid that caused the extinction of the dinosaurs: (a) the asteroid one second before impact; (b) the moment of impact, with an ejecta plume already rising; (c) the "dust pall" in the earth's atmosphere one month after impact; and (d) the impact crater a century after the strike. Interestingly, the artist, Bill Hartmann, is himself a major figure in the history of asteroid impact geology, and was one of the originators of the theory that the earth's moon was formed during a "catastrophic" impact of the earth with a planet-sized body (Hartmann and Davis 1975). Hartmann, an astronomer and geologist, spent his career studying the dynamics of bolide impacts and interpretations of the resulting structural features on Mars and the moon. A prolific illustrator of astronomical scenes, he was intrigued by the Alvarez impact hypothesis when it first appeared. He recalls it as having "seemed reasonable," but explains that after new discoveries he often wondered, "But what would that have *looked like* to a human observer?" (Hartmann, personal communication, April 18, 2019). This series of paintings explored that question. Copyright William K. Hartmann; used with the artist's permission.

A

B

C

D

tion of the catastrophic event presented in the 1980 Alvarez et al. article in *Science*, which concluded by presenting a scenario in which

> an asteroid struck the earth, formed an impact crater, and some of the dust-sized material ejected from the crater reached the stratosphere and was spread around the globe. This dust effectively prevented sunlight from reaching the surface for a period of several years, until the dust settled to earth. Loss of sunlight suppressed photosynthesis, and as a result most food chains collapsed and the extinctions resulted (Alvarez et. al. 1980, 1105).

While it lacks the colorful language that Sagan and others would later use to vividly conjure a nuclear winter, it effectively describes the same phenomenon. In fact, the climate modeling for both studies would be carried out by the same researchers (former students of Sagan's), and in many ways the nuclear winter scenario was directly inspired by the Alvarez hypothesis.

The close association between dinosaur extinction and the potential extinction of humanity through nuclear exchange was thus linked, both scientifically and psychologically, in the extinction imaginary of the 1980s. The appetite for stories about catastrophic extinction—which eventually were extended to hypotheses involving a wandering "death star" responsible for periodic mass extinctions—clearly resonated with a public increasingly anxious about its own potential fate. This was reflected in the often dramatic language used by the press—and even scientists themselves—to describe extinction. As literary scholar Doug Davis has argued, the Alvarez hypothesis "cast the Cold War's nuclear threat into the planet's history. The death of the dinosaurs becomes an atomic war story as researchers across disciplines mobilize the models and metaphors of nuclear war-fighting to read the earth's ancient record of catastrophic impacts" (Davis 2001, 464).

As important as is the connection between dinosaur extinction and nuclear winter in this story, it is only one element of a broad transitional moment in the late-twentieth-century extinction imaginary. The Alvarez hypothesis was a spectacular scientific discovery, but its significance was amplified by a somewhat quieter transformation that took place in paleontology and evolutionary theory in the years on either side of its announcement. Broadly speaking, paleontology went through a "revolution" in methods and agenda that saw a certain approach to the study of life's history—one that was quantitative and theoretical—become established as a subdiscipline known as "paleobiology" (Sepkoski 2012). One of the central themes in this emerging movement was an approach to studying patterns of diversification of life over long periods of time using simplifying models drawn from ecology, and employing computers for sophisticated statistical analysis of large quantities of fossil data. A major result of this study was a new understanding of both mass extinction and the dynamics of ecological and evolutionary stability in geological time.

In its own way, this new understanding had an even greater influence on the scientific and cultural significance of extinction than the Alvarez hypothesis. If the convergence of the Alvarez impact extinction and nuclear winter scenarios represented the apotheosis of the kind of anxiety represented by Cold War nuclear fears, the broader scientific understanding of extinction developed during the 1980s, which increasingly understood mass extinction through the lens of biological diversification, pointed towards a transition towards a new sense of catastrophe: one that would become linked with concerns about humanity's role in upsetting the earth's ecological balance in less sudden though potentially equally devastating ways. That final part of the story will largely occupy the next chapter. For the remainder of this chapter, we will first turn to a brief overview of changes in the paleontological and ecological understanding of diversification and extinction as it developed through the 1970s and early 1980s. Next, we will examine how the Alvarez hypothesis catalyzed what was sometimes referred to as a "new catastrophism" in science, and how this and other well publicized theories, along with the nuclear winter scenario, contributed to a broader extinction imaginary. We will then survey some of the cultural responses to these developments, ranging from postmodernist critiques to apocalyptic film and literature, exploring the diverse ways that catastrophic extinction resonated in late–Cold War "catastrophic" society. Finally, I will argue that scientific and popular interest in mass extinction was in large part responsible for a transition to a new conception of catastrophe, defined in ecological and environmental terms that to a great extent still exists today.

Diversity and Mass Extinction in Life's Past

Writing to a close colleague in 1979, the paleontologist David Raup commented, "I am becoming more and more convinced that the key gap in our thinking for the last 125 years is the nature of extinction" (Raup to Schopf January 28, 1979).[6] What Raup—one of the leading promoters of the "paleobiological" agenda, and one of the foremost theorists of extinction in his generation—meant by this was that the evidence provided by the fossil record about the diversification and extinction of life over time did not support the expectations of the Darwinian view. As his letter went on to explain,

> If we take neo-Darwinian theory at face value, the fossil record makes no sense. That is, if we have a) adaptation through natural selection and/or species selection and b) extinction through competitive replacement or displacement, then we ought to see a variety of features in the fossil record that we do not such as: a) clear evidence of progress, b) decrease in evolutionary rates (both morphologic and taxonomic), c) probably a decrease in diversity.

In other words, Raup was arguing in effect that Darwin's reasoning about selection and extinction was based on assumptions not borne out by evidence. Because Darwin implicitly assumed that Lyell's steady-state model of geology must govern the history of life as well, he assumed that extinction and speciation must balance one another in a slow, gradual pattern. But, observing the state of the field at the end of the 1970s, Raup had reason to challenge this view. Not only had studies of historical patterns in diversity shown that diversification rates had actually *increased* over time; the same studies also suggested that extinctions— mass extinctions, in particular—did not necessarily operate according to normal selective rules. Or, as he put it, "The neo-Darwinian system is at work all the time—producing trilobite eyes and pterosaur flight—

but never really gets anywhere in the long run because the trilobites and pterosaurs get bumped off (through no fault of their own!). . . . The system is always heading toward a steady state but never gets there."

Raup himself had a central role in creating this new understanding of diversification and extinction that had developed over the previous ten years. This was a period of both extraordinary intellectual ferment and debate in paleontology, and rapid advancement in empirical knowledge and techniques for data analysis. Most of the excitement was centered around the study of invertebrates, for the fairly simple fact that—as was discussed in the context of Norman Newell's pioneering work on mass extinction in chapter 4 of this book—the invertebrate fossil record is richer and more complete than the vertebrate record, by several orders of magnitude. Most importantly, by the late twentieth century data collections in invertebrate paleontology had improved to the point where they could be looked upon as reliable sources of evolutionary inferences, and statistical techniques such as multivariate factor analysis had become established as powerful methods for resolving patterns out of the accumulated data. This was the era when computers were adopted as research tools by paleontologists, and Raup was one of the foremost pioneers in this area.[7]

Along with Raup, one of the early innovators in the study of marine invertebrate diversification was the paleoecologist Jim Valentine. Valentine, who has the distinction of being perhaps the only living paleontologist to have belonged to a motorcycle gang (in his 1950s youth), spent most of his career from the early 1960s to the 2000s at the Universities of California at Davis and Berkeley, where he applied many of the exciting developments in theoretical ecology produced by G. Evelyn Hutchinson and his students to the study of the fossil record. In particular, his work helped establish the importance of interactions between ecological and evolutionary hierarchies; in other words, he drew attention to relationships between levels of taxonomic hierarchy (species, genus, family, etc.) and those of ecology (e.g., population, community, ecosystem, biome), and to historical patterns in their changes. For example, his work in the late 1960s and early 1970s argued that different patterns of diversity apply to different levels of taxonomic hierarchy; while diversity was greatest for the highest taxonomic levels (order, class, phylum) very early in the history of life, it has steadily increased at the lower levels (family and genera) over the Phanerozoic eon (the past five hundred-plus million years). Another way of putting this is to say that while the amount of sheer diversity (the absolute number of species and genera) has increased over the history of life, the amount of "disparity," or the degree of difference between groups, has decreased. This is an extremely important point to which we will return in this and the following chapter, but for now it is sufficient to note that Valentine has explained this as a consequence of natural selection settling on particular major body plans and life habits (e.g., phyla), coupled with the increasing crowding of marine ecospace resulting in more and more specialized adaptations to smaller and smaller units of ecological hierarchy.[8]

Valentine's early work raised some significant questions about the mechanics and patterns of marine diversification, and it sparked immediate debate with Raup, who was intrigued by Valentine's approach but wary of potential weaknesses in the data Valentine used. In particular, Raup thought that the appearance of increased diversification at lower taxonomic levels might simply be an artifact of "sampling bias," or the likelihood of particular fossils to be preserved or discovered. Taking potential bias into account, he produced, using statistical analysis, a revised picture of diversification that suggested that diversity had increased to a maximum in the mid-Paleozoic era (roughly three hundred million years ago) and then stabilized at a fairly constant equilibrium up to the present. Raup did, however, acknowledge that (a) the question could not be definitively settled without more and better data at lower taxonomic levels (Valentine's analysis involved extrapolating numbers of species and genera from known data at the family level), and (b) if Valentine's increase was genuine, it would have "broad implications" for our understanding of evolution (Raup 1976, 279–97).

It was recognized at the time that an important question to investigate was whether changes in levels of diversity over time were coordinated between different ecological or taxonomic groups. As discussed in chapter 4, Newell had previously suggested that periods of mass extinction and "exceptional radiation" among coordinated groups of organisms appear to have been a regular feature of the history of life, arguing as early as 1952 that "the rise and fall in apparently evolutionary activity is not at random" (Newell 1952, 385). In 1973, Newell's AMNH colleague John Imbrie (one of the early pioneers in multivariate statistical approaches to paleontological data) and Karl Flessa produced a study that applied statistical tests to Newell's observation that multiple groups of marine organisms appear to have diversified simultaneously at several points in the geologic past. Here they applied factor analysis, which is a statistical test of the probability of meaningful correlation between variables, to determine that most fluctuations of diversity in the fossil record can be correlated into ten "diversity associations" among marine and terrestrial groups. The major finding of their analysis was that while the average rate of taxonomic change for the entire Phanerozoic has been roughly steady, change itself has not been smoothly continuous, but rather can be resolved into a series of distinct and fairly rapid "taxonomic turnovers" or "evolutionary pulsations" where one diversity association gave way to another (Fless and Imbrie 1973).

An important realization developing in the study of diversification was that whether rates of taxonomic change were stable or increasing, underlying mechanisms that controlled when and how these changes occurred were fairly mysterious. As Flessa and Imbrie themselves acknowledged, while their analysis showed a pattern of episodic change, it could not explain it. One way of trying to interpret patterns in diversity over time explored by Valentine, Raup, and others was to incorporate heuristic mathematical models developed by ecologists like Robert MacArthur (who was briefly introduced in the chapter 5), such as the species-area relationship, which predicts the number of species that will occupy a given area or habitat, or the model of island biogeography (developed by MacArthur and E. O. Wilson) that explains the relationship between immigration and extinction on islands as a function of the habitat's size and carrying capacity. For example, Raup's analysis drew implicitly on these heuristic models in its assumption that early diversification would result from expansion into relatively unoccupied ecological niches in the initial evolution of complex life, followed by a dynamic equilibrium once ecospace had been filled. The shape of such a pattern would appear as a "logistic" or S-shaped curve, which is the basic assumption of the MacArthur-Wilson model as well. This heuristic approach was central to the eventual "solution" of the problem of Phanerozoic diversification, but so was the accumulation of better data for statistical analysis.

One of the most important contributors to both aspects of the problem was Raup's colleague—first at the University of Rochester, and then at the University of Chicago—J. John (Jack) Sepkoski Jr., a young paleontologist with a flair for mathematics and a willingness to carry out fairly thankless long-term data collection. As a graduate student at Harvard in the early 1970s, Sepkoski studied with the notable paleontologist and public intellectual Stephen Jay Gould, where he found his true calling (he had originally planned to work on fairly traditional problems in stratigraphy and paleoecology) in the analysis of data about the history of life. A crucial moment came when—as part of a project Gould had initiated with Raup and University of Chicago invertebrate paleontologist Thomas J. M. Schopf, attempting to simulate evolutionary patterns with a computer—Gould asked Sepkoski to begin collecting data on the marine fossil record to test against the simulated outputs. Sepkoski realized that no existing collection or data set was up to the task, so he set about building his own, by drawing on available large compilations as well as obscure monographic literature in several languages.

The initial phase of the project ended up taking a decade, but the result, published in 1982, formed the basis for the first computerized database of marine fossils, and generated a new approach to studying the history of life (though one clearly indebted to earlier studies by Newell, Valentine, Raup, and others).[9] Ultimately, Sepkoski's database would be a central resource for the analysis of mass extinction during the 1980s and beyond, but even before it was completed, Sepkoski himself published several important papers that contributed to the understanding of patterns of diversification. In particular, he was, even more than Raup, drawn to the application of theoretical ecological models to paleontological problems. In 1978, Sepkoski published the first of several papers that argued that general patterns in Phanerozoic marine diversity could be modeled as a logistic curve—following MacArthur and Wilson—in which an initial phase of rapid diversification was followed by leveling to equilibrium.[10] What distinguished Sepkoski's analysis from earlier attempts (such as those of Raup, Imbrie, and Flessa) was both the improved quality of the data (drawn from his unpublished database) and the attempt at deriving causal explanations for the phenomenon. One conclusion of Sepkoski's 1978 paper was that major diversification events, such as the Cambrian "explosion" of multicellular life, can be explained simply by the dynamics of internal ecological relationships. That is, given a large, fairly unpopulated ecosystem, evolutionary radiation from relatively few, unspecialized groups to many more specialized ones is the logical expectation, and it accounts for the initial exponential portion of the logistic curve. The second phase, where the graph levels off to equilibrium, is likewise explained by so-called density-dependent factors; as ecospace becomes filled, greater competition, smaller population sizes, and reduced niche sizes will act as a natural curb on further diversity increases.

Complicating Sepkoski's initial study, however, was his realization (as a result of the improved resolution his new database was making available) that Phanerozoic diversification was a matter not of a single logistic pattern but of multiple, overlapping, successive ones, each characterized by different constituent "faunas," and each carrying its own evolutionary trajectory. In 1979 Sepkoski updated his analysis by identifying a second, post-Cambrian logistic pattern that "greatly altered both diversity and faunal composition in the world ocean" (Sepkoski 1978, 223–51). Not only was this pattern of diversification characterized by an entirely different collection of organisms—he called it the "Paleozoic shelly fauna" because mollusks came to dominate the seas—but it also appeared to be correlated with the decline, to extinction or low levels of diversity, of the previous Cambrian fauna. To this two-curve model Sepkoski would in 1984 add a third diversification curve, beginning in the Triassic (around two hundred million years ago) and characterized by vertebrates and larger crustaceans, which has continued its upward sweep through the present. This completed what came to be known as the "kinetic model" (because diversity "moves" up and down the successive logistic curves) or simply the "Sepkoski curve" (fig. 5.2; Sepkoski 1978, 246–67).

A significant conclusion of Sepkoski's work on evolutionary faunas was his emphasis on the role of "internal" constraints—the dynamics of populations under ecological pressures—for constraining patterns of

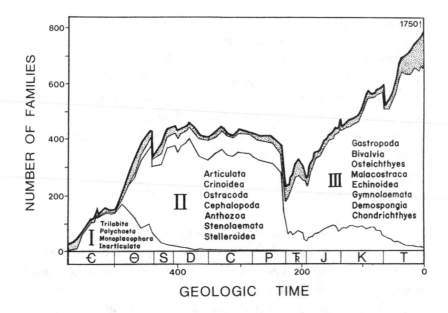

FIGURE 5.2 Jack Sepkoski's graph of three successive "faunal stages" in the diversification of life over time (Sepkoski 1981, 49). The sharp "dips" in diversification show the "Big Five" mass extinction events as losses in standing diversity.

diversification. However, if normal patterns were generated by these internal constraints, this provoked the question of why major faunal turnovers, which appeared to interrupt and then restart the normal logistic pattern, took place. In fact, the initial (single-curve) logistic model would appear to support the Darwinian assumption of a steady-state equilibrium in the history of life, and furthermore would appear to be nicely explained by classic Darwinian mechanisms like competition and selection. This is precisely what Raup, writing to Schopf at exactly the time Sepkoski was publishing his results, was arguing had been thrown into doubt. What explains the discrepancy?

If Sepkoski had only ever proposed the single logistic curve, then Raup might have had little basis for his assertions to Schopf. However, Raup was working closely with Sepkoski and was intimately aware that Sepkoski's thinking about diversification had changed significantly between 1978 and 1979. In identifying a second, and eventually a third, logistic pattern, Sepkoski found his attention drawn to irregularities in the curves his data described. While his idealized model represented the logistic pattern as a smooth, unblemished line, the actual pattern was anything but. In fact, the three-curve diagram (fig. 5.2) shows a number of quite prominent spikes and troughs in the pattern—several of which appear to be too steep to be merely the result of random fluctuations. The most significant of these, in a statistical as well as a colloquial sense, appear not coincidentally at the boundaries between major geological periods—which, we will recall, were initially identified by Cuvier and others because they were demarcated by major changes in the fossils populating corresponding stratigraphic layers. This was, of course, a major reason for Cuvier's theory of global catastrophes, and the observation of similar deviations in data patterns was the basis for Newell's assertion of the reality of mass extinctions. A similar conclusion appeared obvious to Sepkoski: the major dips, which could not be explained as arising from poor data or random inflections, represented major periods of mass extinction.

It was significant enough that Sepkoski's higher-quality data seemed to confirm Newell's earlier proposals; what made the results even more important was what they implied in the context of diversification patterns. Each of the five major troughs (labeled in the figure) represents a well-known period of suspected major extinction—that much was already known from anecdotal evidence from the fossils. But when viewed in the broader context of faunal replacement, it becomes apparent that, following the most dramatic extinction events—such as the end-Permian extinction, where the number of living families dropped almost 50 percent—came periods of evolutionary radiation that introduced distinctively new forms of life and were dominated by entirely different organisms. Sepkoski was able to model these events as "perturbations" of the logistic pattern, after which a new logistic growth phase would begin again. The inescapable conclusion was that not only did mass extinctions feature as fairly regular episodes in life's history, appear to affect "associations" (in Flessa and Imbrie's terms) of seemingly unrelated groups, and seem to open opportunities for previously marginal groups (as the mammals following the K-T trough), but they also appeared to have a major role in permanently altering the subsequent history of the diversity of life, in most cases by actually promoting *greater* subsequent rates of diversification. The implication was that without mass extinctions, the earth might still be populated by those forms that radiated during the Cambrian. In this context, mass extinctions appear to be one of the major engines of diversification in life's history.

Of course major questions remained, including the possible mechanism(s) that could produce this pattern. But, at the beginning of the 1980s, research on diversification had reopened debates that had not been broadly entertained for well over a century. And an irony was that, as Sepkoski later reflected, "the data were not even compiled to investigate mass extinction, and only serendipitously contained information useful in the [subsequent] extinction controversies" (Sepkoski 1994, 132).

Bad Genes or Bad Luck?

One of the most intriguing potential revelations of the studies of diversification described above was the suggestion that extinction, and thus evolution, might not always be directed by Darwinian rules of selection and adaptation, at least as these concepts had been traditionally understood. This is what Raup was hinting at in his 1979 letter to Schopf: that, as he put it, trilobites and pterosaurs died out "through no fault of their own," raising the possibility that "extinction is random with respect to fitness" (Raup to Schopf, January 28, 1979). This is a question that had been occupying Raup for a number of years, dating back to his collaboration on computer simulations of evolutionary patterns, and it would ultimately inform a radically new understanding of extinction that he and colleagues would develop throughout the 1980s and 1990s.

As Raup pointed out in a presidential address to the Paleontological Society in late 1977 titled "Approaches to the Extinction Problem," virtually every explanation of major episodes of extinction that had ever been proposed had assumed that extinction must result from some "common failure or deficiency" among affected species. Illustrating the point with the example of the survival of the echinoids (urchins) and the extinction of the blastoids (a previously well established related group) at the end of the Permian, Raup observed, "We assume that 'echinoidness' was somehow better than 'blastoidness'" (Raup 1978, 517–18). But why should this necessarily be the case? This kind of reasoning, he argued several years later, at best produces unresolvable debates about which traits were to blame, and at worst leads to tautological reasoning. In a case of differential survival and extinction of two contemporary groups, there is usually very little "evidence at all of the inferiority of the victims *except* their lack of survival." Raup considered this kind of thinking to result from "overuse of the Darwinian paradigm," since it was the assumption that extinction must result from inferiority that became the evidence of that inferiority—a circular argument.[11] The appeal of this reasoning seemed obvious to Raup: it implied that extinction has a kind of moral, and that evolution is "a fair game where goodness triumphs in the end," a lesson he suspected "fits well with the traditional Calvinist views that many of us grew up with."

The problem with Raup's alternative view of extinction was that, in the late 1970s, there was very little evidence of a phenomenon that could provide a mechanism for potentially nonselective mass extinctions. But unbeknownst to Raup, that would soon change. In 1978 the geologist Walter Alvarez and his father, the Nobel Prize–winning physicist Luis Alvarez, had embarked on a project that they hoped would combine their respective fields of expertise: they suspected that it might be possible to more reliably correlate stratigraphic layers across different geographical locations by identifying distinctive levels of trace elements present in multiple locations—a project that nicely combined the techniques of a nuclear physicist and a geologist. It was this, and not any attempt to solve the riddle of dinosaur extinction, that led them to well preserved sections of the K-T boundary in Gubbio, Italy, where they detected unusually high levels of iridium (one of the elements they identified as a potential stratigraphic signature) precisely at the location where the dinosaurs disappear from the fossil record. Puzzled by this anomaly, they spent nearly two years carefully analyzing their findings, and ultimately submitted an unusually long paper to *Science* titled "Extraterrestrial Cause for the Cretaceous-Tertiary Extinction" (Alvarez et al. 1980).

The publication of the initial Alvarez paper in June 1980 would be the beginning of a series of rapid developments in extinction theory that gained enormous popular attention, but initially the paper caused few ripples in the public eye. For example, the *New York Times*, which would go on to extensively cover developments in extinction over the next decade, mentioned the paper only once that year, in a short article reporting on "two new theories" of mass extinction. Of these theories (the other was a theory of lunar volcanic eruption published in *Nature* the same year), the Alvarez hypothesis received just five scant paragraphs, less than a quarter of the article's length, and the article gave little indication of the potential importance of the proposal for understanding the general dynamics of extinction.[12]

Among geologists and paleontologists, however, it was another matter. The Alvarez paper attracted immediate attention and controversy, and led to an almost instantaneous consideration of its various proposals and their merits. In October of 1981, a conference titled "Large Body Impacts and Terrestrial Evolution" was convened at the Snowbird ski resort in Utah, jointly sponsored by the National Academy of Science and the Lunar and Planetary Institute. This conference, which attracted more than a hundred participants to hear more than fifty papers, was billed explicitly as having "grow[n] out of the stimulation of the scientific community by the provocative paper by Alvarez, et al. (Science, 1980, 208, p. 1095) on the significance of iridium geochemical anomalies at the Cretaceous-Tertiary boundary."[13] In addition to an update from the Alvarez team, the published proceedings featured articles by most of the major paleontologists involved in studying extinction, including Newell, Raup, Sepkoski, Schopf, Dale Russell, William Clemens, and others. In his introduction to the volume, Leon T. Silver rather breathlessly proclaimed, "Catastrophism has been rekindled!" and opined, "Among students of the evolutionary paths, the ecological sensitivities, and the causes of extinction in various classes of biota, the Alvarez and others [*sic*] proposals have shed new light for some, but have drawn some heat and not a little smoke from others" (Silver and Schultz 1982, xiii).

Indeed, this meeting seems to have been the real launching point for wider interest in mass extinction for two primary reasons. The first is that the conference was attended by several journalists, whose interests were finally piqued enough to begin penning sometimes sensationalist accounts of the scientists' findings. The second was that the meeting put the Alvarez team in direct contact with Raup and Sepkoski, and effectively drew the independent threads of research—by the paleobiologists on diversification and mass extinction, and by the geophysicists and astrophysicists on impact evidence and physical consequences—under the same umbrella. In some ways the two groups were natural allies; paleobiologists like Raup and Sepkoski had continued to face considerable skepticism towards their accounts of mass extinctions from paleontologist and geologist colleagues who viewed "catastrophism" with suspicion, while the Alvarezes faced marginalization as "outsiders" by biologists and paleontologists. Years later, Walter Alvarez reported having been "delighted" that Raup and Sepkoski were willing to talk to him at the Snowbird meeting and to give his ideas an open hearing.[14]

In terms of furthering the understanding of past mass extinctions, the 1981 conference also saw two important new pieces of information offered. The first was a paper by Brian Toon and several colleagues offering an estimate and simulation of the atmospheric consequences of a ten-kilometer asteroid impact, based on comparisons with both observed volcanic eruptions and computer climate models. This was an important study because the Alvarez hypothesis required significant atmospheric and environmental consequences to have triggered the widespread mass extinctions observed in the fossil record; the impact alone would likely have caused only local extinctions. Here the paper by Toon et al. was favorable, estimating that the energy produced by the impact would have been on the order of a million times that of the 1883 Krakatoa eruption, disrupting global photosynthesis for at least three months and causing freezing temperatures for half a year.[15] For some, these calculations bolstered the likelihood of the Alvarez hypothesis, but others worried that the atmospheric effects weren't sufficiently severe. In the eyes of the most skeptical scientists, the hypothesis would only be vindicated when an impact crater of appropriate size dating to the right period was discovered—which would not happen until the 1990s.

The second major paper was presented by Sepkoski on continuing analysis he and Raup had performed on their fossil database in an attempt to test whether mass extinctions that had been qualitatively identified by past researchers would appear as statistically significant events using Sepkoski's new data on marine families. In a companion paper published the same year (1982) in *Science*, Raup and Sepkoski explained that their procedure was to calculate the number of family extinctions that had taken place during each of the seventy-six recognized stratigraphic stages since the early Cambrian—essentially by counting the last appearances of families in Sepkoski's database and correlating them with time. Next they performed regression analysis on the seventy-six extinction points, to identify statistical outliers. Five such points appeared, at exactly the locations where previous qualitative analysis had suggested that mass extinctions may have taken place. These included the greatest extinction in the history of life, at the end of the Permian, when between 77 and 96 percent of all marine species may have died out, but also four major additional events—each representing a moment when at least 15 to 22 percent of standing familial diversity had been lost over less than fifteen million years—including the K-T event.[16] The upshot of this research was both to validate the so-called "big five" mass extinctions as statistically significant empirical phenomena, and to demonstrate that "major mass extinctions are far more distinct from background extinction than has been indicated by previous analyses" (Raup and Sepkoski 1982, 1502).

Despite cautioning at the end of their article in *Science* that "the data do not tell us, of course, what stresses caused the mass extinctions," Raup and Sepkoski, as well as many other observers, found the Alvarez hypothesis to offer a compelling potential mechanism that might be generalized beyond just the K-T event. In a long piece in the *New York Times* based on coverage of the 1981 Snowbird meeting, the journalist Walter Sullivan, who had authored the tepid earlier article describing the Alvarez hypothesis, reported: "From evidence still being gathered, a number of geologists believe the Cretaceous extinction that killed off the dinosaurs and numerous other animals 65 million years ago coincided with the impact of an asteroid or giant meteorite. Perhaps, they say, such catastrophes caused most of the great extinctions and are bound to happen again" (Sullivan 1982). Sullivan also observed, somewhat inaccurately, that the current debate "has its roots in the 19th century when those who, reared in a Biblical tradition, believed in catastrophic events were pitted against adherents of gradual change"; but he noted, "Today many geologists have adopted a mix of both concepts."

The Snowbird meeting—and the reactions of Sullivan and other participants and observers—raises an important point: acceptance of the Alvarez iridium data, and even the interpretation that it resulted from a major bolide impact, did not necessarily require adopting a "catastrophist" viewpoint. As Raup put it in a summary of the conference published in the journal *Paleobiology*, the updated Alvarez evidence had led even those "not inclined toward catastrophic explanation" to agree that "there was probably a large body impact at the end of the Cretaceous" (Raup 1982, 1). But it was one thing to acknowledge the probability of an asteroid striking the earth now and again; after all, the moon gives vivid testimony that such events have happened in the past. It was another to accept—as only some of the Snowbird participants did—that the impact was solely responsible for the extinction of the dinosaurs and other groups associated with the K-T event. A number of other candidate explanations still circulated, the most serious of which was concurrent increased volcanic activity in a large area of what is now western India. And while some found the impact scenario promising, very few people were willing to suggest that such extraterrestrial impacts might have been responsible for *all* of the mass extinctions identified by Raup and Sepkoski. As Raup himself summarized it, "When all the returns are in . . . we may have little more than an important new tool for stratigraphic correlation (many Ir spikes scattered throughout the record) *or* we may have totally new paradigms for geological and paleobiological interpretation of the Phanerozoic" (Raup 1981, 1–3). Privately, in a letter to Schopf, he was somewhat more effusive, writing that "this is one really exciting time to be alive in geology and evolutionary biology. We may be witnessing a major revolution—or perhaps a large red herring" (Raup to Schopf, October 25, 1981).

For the developing theories to genuinely be considered catastrophism, a few important facts would have to be established. First, that extinction events involved multiple unrelated groups in a wide variety of global environments, and were of sufficient magnitude that they would stand out recognizably from normal background rates. Second, that

such events were a *regular* feature of the history of life and not just iso-
lated aberrations. Third, that they were triggered by some mechanism
external to normal evolutionary and geological processes—that is to
say, that they involved mechanisms that were not reducible to normal
evolutionary dynamics (e.g., competition and selection between individuals) or observable environmental processes (gradual climate or sea
level change, etc.). Fourth, that the consequences of these events—on
future evolution and diversification or environmental composition or
both—were distinctive, lasting, and perhaps permanent. At the beginning of 1982, when Sullivan's article appeared, there was tantalizing evidence for all four criteria, but little consensus. Points one and two were
addressed by Sepkoski and Raup's analysis of extinction data, but not
definitively enough to erase concerns about the nature of their data and
analysis (for example, whether data collected for families was legiti-
mately extrapolated to rates of species extinction, or whether their time
series was fine-grained enough to reliably detect rapid events). Point
three appeared to have been more clearly resolved by the Alvarez data,
but the broader hypothesis—that the impact would have triggered a
global environmental crisis sufficient to have caused mass extinction—
was less certain. And the final point—the impact of extinction on the
future of life—had hardly yet been explored.

The New Catastrophism and the Cold War

I argue that for the research on diversification and extinction carried
out from the early 1970s to the mid-1980s to genuinely deserve the label
"new catastrophism"—and I believe that it does—we need to consider
broader developments in science and culture, as well as arguments
made by paleontologists and geologists. From the very beginning in
the nineteenth century, when William Whewell coined the terms, labels
like "catastrophism" and "uniformitarianism" were much more than descriptions of scientific positions in geology. They were, in the first instance, characterizations of scientific methodologies. So-called catastrophists (whose ostensible proponents almost never used the term
themselves) were alleged to be wildly speculative, perhaps even theo-

logically motivated theorists, while uniformitarians were understood to be practitioners of sober, empirically sound, and inductively reasoned science. More than that, though, as discussed in chapter 1 of this book, catastrophism was often associated with a view of history, including human events, that saw revolution and upheaval as part of the regular and perhaps even desirable course of things, while uniformitarianism reflected a belief in stability, constancy, and predictable order. Especially during the volatile years of the mid-nineteenth century, these two labels had strong cultural and political valences, and reflected competing extinction imaginaries.

If this was the case in the nineteenth century, then why not in the twentieth? It is certainly the case that the first level of association continued to adhere to the two terms; after all, in geology one of the major objections to the geological theory of plate tectonics was that it violated the principle of uniformitarianism.[17] Likewise, in biology the Modern Evolutionary Synthesis of the 1940s had stamped out any suggestion that evolutionary change could happen in any way other than the slow, minute accumulation of beneficial traits in populations over very long periods of time. In the evolutionary and environmental sciences, even in the 1980s and beyond, the assertion of abrupt, significant, and discontinuous change remained anathema in many corners.

On the other hand, scientific and cultural forces had been brewing for some time that likely prepared the way for the return of a catastrophic view of extinction. Some of these—nuclear fears, environmental anxieties, political instabilities, and war—have already been discussed in previous chapters. Others, such as the radical politics of the late 1960s and 1970s (student activism, antiwar protests, civil rights, women's and gay rights movements) gave the period a sense of impending transformation, rupture, and even revolution that colored perceptions of scientists and public alike. It is worth noting—although difficult to make firm causal claims about—the fact that a number of participants in the mass extinction debates had taken part in 1970s-era protest movements as students or young faculty, and had quite actively participated in a counterculture whose slogans—"The Times They Are a-Changin'," "Anarchy in the UK"—betokened and celebrated radical change in politics and society.[18]

The political landscape of the 1980s was certainly eventful, and particular developments had an influence on a developing culture of catastrophic anxiety that was not just a continuation but the apotheosis of nuclear fears discussed in the previous chapter. In fact, in many ways the period from the late 1960s through the beginning of the 1980s saw something of a reduction in general social anxiety about the threat of nuclear war. Following the Cuban Missile Crisis, it appeared that both superpowers had stepped back from the brink, and genuine diplomatic efforts aimed at reducing the potential for nuclear catastrophe—underlined by the Strategic Arms Limitation Talks (SALT I and II) carried out throughout the 1970s—seemed to promise a genuine reduction in the capacity for human civilization to destroy itself. This is not to say that the 1970s were a peaceful decade, but rather that the focus of social unease, especially in the United States, was diverted to a variety of environmental, economic, and domestic political concerns.

A combination of circumstances, however, arose at the beginning of the 1980s that would elevate cultural anxieties around nuclear conflict to a new peak. Two events in late 1979—the Iran hostage crisis and the Soviet invasion of Afghanistan—dramatically raised tensions in international politics, and the election of President Ronald Reagan in 1980 brought an escalation of Cold War posturing that had been scaled back during the previous administration of Jimmy Carter. During his first term in office, Reagan took a number of steps that contributed to a much more aggressive stance towards the Soviet Union, such as dramatically increasing US armed forces deployment, developing the MX missile program (explicitly designed to counterattack a Soviet first strike), deploying the Pershing II missile system in West Germany, and proposing the Strategic Defense Initiative (or "Star Wars") missile defense system to potentially repel a Soviet attack. Additionally, Reagan's rhetoric ratcheted up hostility toward the Soviet Union to levels not seen in over a decade, including statements that communism would wind up "on the ash heap of history," and that the Soviet Union was an "evil empire."[19]

It was into an already charged atmosphere of anxiety and uncertainty, then, that Carl Sagan dropped his dire assessment that the consequences of even a limited nuclear exchange could have effects far beyond those previously imagined, almost surely leading to the destruction of civilization and perhaps the extinction of the human species. The public announcement came in a short piece Sagan published on October 30, 1983, in *Parade* magazine (a syndicated supplement then carried by many US newspapers in their Sunday editions) titled, simply, "The Nuclear Winter." Ominously, it warned that while a nuclear war "would represent by far the greatest disaster in the history of the human species and, with no other adverse effects would probably be enough to reduce at least the Northern Hemisphere to a state of prolonged agony and barbarism. . . . the real situation would be much worse" (Sagan 1983c). Specifically, Sagan explained that the dust and soot from the explosions and resulting firestorms would result in a prolonged period of cold temperatures and darkness that would "represent a severe assault on our civilization and our species." There was, he argued, no question of a "winnable" nuclear war, even one involving a relatively moderate exchange. The only real question would be the extent of the catastrophe. As Sagan explained,

> Many biologists, considering the nuclear winter that these calculations describe, believe they carry somber implications for life on Earth. Many species of plants and animals would become extinct. Vast numbers of surviving humans would starve to death. The delicate ecological relations that bind together organisms on Earth in a fabric of mutual dependency would be torn, perhaps irreparably. There is little question that our global civilization would be destroyed. . . . And there seems to be a real possibility of the extinction of the human species.

This statement, coming in stark terms from one of the most recognizable and trusted scientists in America, who had been a regular visitor to living rooms through his wildly popular television series *Cosmos*, came as a shock to many. North Americans and Europeans had for many years lived in the shadow of the mushroom cloud, but Sagan's warning presented the threat as something potentially worse: a major extinction event.

From the very beginning, Sagan and his colleagues explicitly presented the nuclear winter scenario as analogous to recent discoveries about the major mass extinctions of life's past. In two much more detailed reports on the climactic and biological consequences of nuclear exchange published in December of 1983 in *Science*, little doubt was left as to the potential severity of the aftermath. "A severe extinction event could ensue," one article warned, "leaving a highly modified and depauperate Earth" in which "species extinction could be expected for most tropical plants and animals, and for most terrestrial vertebrates of north temperate regions, a large number of plants, and numerous freshwater and some marine organisms" (Ehrlich et al. 1983, 1299). The same article highlighted the importance of ecosystems that provide "food and their maintenance of a vast library of species from which *Homo sapiens* has already drawn the basis of civilization," and stressed that the "loss of these genetic resources would be one of the most serious potential consequences of nuclear war" (Ehrlich et al. 1983, 1298). In this respect, the effects of war were described explicitly in terms of a loss of diversity that could have "irreversible" consequences for the future global ecosystem. This was precisely the way paleontologists had come to understand the aftermath of major mass extinction events, and the accompanying article, on atmospheric effects, explained, "The discovery that dense clouds of soil particles may have played a role in past mass extinctions of life on Earth has encouraged the reconsideration of nuclear war effects" (Ehrlich et al. 1983, 1293–1300).

Another feature of the presentation of the nuclear winter scenario was the explicit mobilization of a large group of scientists representing many disciplines to make a public call for attention and action. These efforts were somewhat reminiscent of actions taken by scientific organizations and advisory panels in the past on matters of public concern such as nuclear fallout or environmental disaster, but the 1980s saw new levels of scale and media savvy in such activities. Many of the scientific articles themselves were coauthored by large teams: for example, Sagan and his colleagues Richard Turco, Brian Toon, Tom Ackerman, and James Pollack (often abbreviated as "TTAPS," from the authors' last names) published a number of articles in a variety of disciplinary journals, and the 1983 *Science* article on biological consequences of war was signed by some twenty authors including Paul Ehrlich, Sagan, the evolutionary biologists Peter Raven and Ernst Mayr, the ecologists Robert

May and Norman Myers, and Stephen Jay Gould. Moreover, Sagan's *Parade* article was followed by a major scientific conference that took place in Washington on October 31, 1983 (the day after the magazine piece appeared), which was attended by five hundred scientists and one hundred journalists, and remarkably featured a ninety-minute discussion with Soviet counterparts via live satellite link. This kind of major mobilization and presentation as scientific extravaganza — complete with press releases and mass media coverage, and followed up with coordinated editorials, articles, and books aimed at popular audiences — would become a feature of future efforts to raise public awareness about biodiversity loss and global climate change. From the 1980s onward, at least as far as topics relating to environmental and biological catastrophe were concerned, the line between scientific and public discourse would effectively be erased.

The 1983 meeting on nuclear winter certainly brought a spate of media attention, with notice in opinion pieces and editorials in most of the major news outlets in the United States and many abroad. A distilled account of the proceedings was published the following year by W. W. Norton as a book titled *The Cold and the Dark: The World after Nuclear War* (fig. 5.3). Described in the jacket copy as "a work of science, not science fiction," the book featured two chapters on atmospheric and biological consequences of nuclear war that were authored by Sagan and Ehrlich respectively, followed by transcripts of panel discussions and the conversation with Soviet scientists. While the two main chapters largely covered the same material as the more technical articles published the previous December in *Science*, the foreword and introduction to the volume turned up the rhetoric about potential extinction and the connection between nuclear winter and prehistoric impact scenarios.

Interestingly, the foreword and the introduction were written not by main researchers in the study, but by academic administrators of high standing whose perspectives presumably were meant to bridge between the general reading public and the experts analyzing the data. In his foreword, Lewis Thomas, chancellor of the Memorial Sloan-Kettering Cancer Center in New York, gave a passionate call to avert a catastrophe that could "mean nothing less than the extinction of much of the

FIGURE 5.3 The cover of Sagan and Ehrlich's study of the nuclear winter phenomenon, *The Cold and the Dark*. Reprinted with permission of W. W. Norton & Co.

Earth's biosphere," potentially reducing the planet's life to "a level comparable to what was here a billion years ago" (Thomas, in Ehrlich et al. 1984, xxi–xxiii).[20] Thomas went on to note, "The last great extinction of planetary life occurred 65 million years ago . . . probably as a result of an asteroid collision with the Earth. It is this kind of event that is forecast by the models used in these studies." In the book's introduction, the Stanford University president, biologist, and former FDA commissioner Donald Kennedy went even further in making his connection, arguing, "This new view results in part from a new general paradigm in scientific thinking about the processes that have influenced Earth's history"—one that overturned Lyell's "revolution" against "the catastrophist view," in favor of "one based upon a doctrine of uniformitarianism":

> Today the earth sciences are in the middle of a second revolution, triggered by the remarkable discoveries of plate tectonics, and the emphasis has moved back toward more dramatic events. Increasingly, it is recognized that major discontinuous interventions such as volcanic eruptions and asteroid collisions may have had profound effects on the history of the Earth and of the life on it (Kennedy, in Ehrlich et al. 1984, xxx).[21]

One important point to make here is the obvious—and acknowledged—role that the study of mass extinctions in the geological past had on the construction of the nuclear winter scenario. This was broadly evident in how the biological consequences of nuclear war were understood—as disruptions in a stable ecological equilibrium, with potentially permanent consequences for biological diversity—as well as in the atmospheric and climatic models that were presented for the winter itself. Indeed, the central team of climate modelers for the nuclear winter scenario was precisely the same group of scientists—"TTAPS," minus Sagan—who had developed the atmospheric model for the Alvarez hypothesis presented at the 1982 Snowbird extinction meeting and in several subsequent papers. The connection between nuclear winter and the Alvarez hypothesis was frequently referred to in press accounts, where the Alvarez study was variously described as having "inspired," "helped to shape," been "similar to," or been a "variant on" the nuclear winter scenario.[22] Little surprise, then, that in the public eye, at least, the two scenarios were often easily conflated.

Nemesis and a New View of Life's History

As significant as the Alvarez impact hypothesis was on popular and scientific perceptions of extinction in the 1980s, it was only one piece of a larger suite of extinction research during that decade that attracted wide notoriety and excitement. In fact, following the initial studies by the Alvarez group, Raup and Sepkoski, and others, the study of mass extinction and its consequences very quickly became something of a cottage industry within paleontology and related fields. While some commentators observed wryly that the hubbub surrounding extinctions had made "media stars out of some of our colleagues" (feature articles in *Time, Newsweek, Discover*, and other publications of this period often featured awkward photographs of paleontologists crouched over computer printouts), it was certainly the case, as Karl Flessa commented in a 1986 opinion piece in *Paleobiology*, that extinctions were "IN" (Flessa 1986, 329).

As discussed previously, Raup and Sepkoski's quantitative analysis of diversification had by 1982 identified five major mass extinctions that rose far above the threshold of background noise, along with a dozen or so more "minor" events of more questionable statistical relevance. One of the questions their initial article had raised was about the timing and regularity of these extinctions: did they distribute randomly in time, or was there some regular pattern? To answer this question, Raup and Sepkoski analyzed a refined subset of the data from their original study—removing data from poorly sampled or otherwise questionable groups—and performed what Raup later described as a kind of "gestalt" experiment, where they generated a long printout and stood across the room, looking for patterns.[23] Sepkoski remembered the graph looking like "a seismic reflection profile," with peaks regularly spaced in time. "'Oh shit,'" Sepkoski recalled thinking; "'Fischer and Arthur were right'" (Sepkoski 1994, 143).

Sepkoski was referring to a 1977 paper by the Princeton geologist

Alfred Fischer and his student, Michael Arthur, who had proposed a thirty-two-million-year cycle in episodes of marine diversification. According to Fischer and Arthur, fluctuations in the "rhythm" of diversity in the oceans—between periods of higher and lower diversity—correspond to cycles of change in ocean temperatures, sedimentation rates, and general climate patterns. While they did not directly attribute a cause to this pattern, they suggested corresponding cycles of change in solar activity or plate tectonics as likely agents. And although they did not propose this as a "catastrophic" process, they nonetheless acknowledged that these were "changes of a sort not considered by Hutton, Lyell, and other classical uniformitarianists [*sic*]" (Fischer and Arthur 1977, 25, 19).

While Fischer and Arthur's paper had gone relatively unnoticed, Raup and Sepkoski's analysis, which they quickly wrote up for publication as a short article in *Proceedings of the National Academy of Sciences* in early 1984, attracted immediate attention and controversy. This was in part due, no doubt, to their use of a more refined data set than Fischer and Arthur had, and to their application of rigorous statistical analysis to their results, something Fischer and Arthur had not done. More broadly, the differential reaction to these two studies shows the major shift that had taken place in the field in just a few years: the combination of the Alvarez team's impact hypothesis and paleontological analysis of mass extinction events had introduced a climate in which it was now legitimate, if not universally accepted, to at least discuss the possibility of a "new catastrophism."

What Raup and Sepkoski found was a somewhat shorter cycle in which mass extinctions resolved as statistically significant peaks at almost precisely twenty-six-million-year intervals. Raup and Sepkoski did not themselves invoke the term "catastrophism," and their initial article struck a fairly cautious tone, focusing on the data and on the robustness of their statistical tests, rather than on the broader consequences of the pattern. Nonetheless, they acknowledged at the end of their paper that if the twenty-six-million-year periodicity held up, then "the implications are broad and fundamental" (Raup and Sepkoski 1984, 805). While the authors acknowledged that the "forcing agent" behind such a cycle remained mysterious, they expressed their preference for "extra-terrestrial causes . . . where the cycles are of fixed length and measured on a time scale of tens of millions of years." But they concluded that, regardless of the exact mechanism, "the implications of periodicity for evolutionary biology are profound," since they suggested that the "evo-lutionary system" may be significantly directed by external "perturba-tions" without which "the general course of macroevolution could have been very different."

Interestingly, it was only after the presentation of Raup and Sep-koski's twenty-six-million-year periodicity study that the press began to take an interest in extinction science in earnest. While some notice had been taken of the Alvarez scenario, from late 1983 onward a slew of articles appeared in major international newspapers, as well as in popular science and news magazines (fig. 5.4). The sociologist Eliza-beth Clemens has tallied up the number of magazine articles mention-ing impact debates during the 1980s, and the results are intriguing. Between 1980 and 1982 (the three years following the Alvarez team's announcement), the thirteen major magazines she surveyed—ranging from *Time, Newsweek,* and the *New Yorker* to *Discover, Scientific Ameri-can,* and *Science Digest*—published a combined thirty-two articles on mass extinction theories. However, over the next three years—between 1983 and 1985—that total grew to a combined fifty-nine, and the up-ward trend continued for several more years.[24] Clemens's analysis does not include newspaper articles, but there my own more qualitative sur-vey reveals an even more striking pattern: whereas very few newspapers gave the Alvarez discovery more than passing initial mention, the topic suddenly became prominent after the periodicity hypothesis became widely known.

My explanation for this phenomenon has to do with timing and convergence. The Alvarez scenario, while intriguing to those mem-bers of the public with an interest in dinosaur extinction, did not an-nounce itself as a major revision of conventional scientific understand-ing. It merely accounted for a fact already fairly well accepted—that the dinosaurs died out suddenly—albeit with a fairly dramatic potential mechanism. None of the Alvarez group's published scientific accounts

FIGURE 5.4 *Time* magazine cover with the headline "Did Comets Kill the Dinosaurs?" (May 6, 1985).

featured the vivid language that would later be used to describe the scenario; that would come later—for example, in a number of popular books, including Walter Alvarez's own *T-Rex and the Crater of Doom*, that described the fatal impact in language reminiscent of atomic war literature. And while the story of the death of the dinosaurs was potentially fertile ground for analogy and moral lessons regarding the fate of humanity, it appears that such connections became apparent only later, after other developments came to be registered in the public consciousness.

Though published in early 1984, the Raup-Sepkoski periodicity analysis was in fact first presented in August 1983 in Flagstaff, Arizona, at a conference called "Dynamics of Extinction," which featured papers discussing not only the mass extinctions of the geological past but also potential extinctions in the present. Paul Ehrlich, for example, gave a talk in which he warned both of the threat to "the very future of humanity" caused by biodiversity loss (more on this topic later), and of "the single greatest threat of extinction hanging over the planet—large-scale thermonuclear war" (Ehrlich 1986, 162). The ecologist Daniel Simberloff likewise described current levels of extinction in tropical regions as an "imminent catastrophe," and the paleontologist David Jablonski gave an early exposition of his developing work—expanding on Raup's studies—on the ways in which mass extinctions shape the course of evolution by changing the "rules" of selectivity during times of crisis.[25]

The Flagstaff conference was attended by journalists, and accounts of the meeting were eventually published. What is particularly interesting, though, is that some of the most prominent descriptions, focusing especially on the extinction periodicity hypothesis, were published only several months later. For example, John Noble Wilford's piece in the *New York Times*, "Study Indicates Extinctions Strike in Regular Intervals," appeared on December 11, 1983, and drew attention to Raup and Sepkoski's "potentially revolutionary" finding that "elevates the importance of rare, catastrophic events in setting the course of life," ultimately "pushing science further in accepting catastrophe as a 'normal' part of the earth's history" (Wilford 1983). In similar terms, in a *MacLean's* article published on December 26, a journalist opined that if periodicity was confirmed, "the way in which mankind views the evolution of life on Earth may change irrevocably"; the article concluded that "human beings can ultimately thank the jolt of an errant asteroid, or some cyclical extraterrestrial event, for allowing them to step into their role as Earth's most intelligent creatures" (Ohendorf 1983).

The significance of the timing of these articles, of course, was that they appeared *after* both the airing of *The Day After* and the announcement in late October of the nuclear winter hypothesis. It is certainly conceivable that production schedules or other assignments delayed the accounts of the Flagstaff conference, or that the periodicity hypothesis would have received the same attention independently; but I think the broader phenomenon, that the public took special notice of mass extinction only after late October 1983, is more than mere coincidence. In hindsight, what appears to have happened is that the Alvarez discovery stoked some mild initial interest that did not translate to broader public fascination until the catastrophic death of the dinosaurs had been placed in a context that spoke meaningfully to contemporary anxieties (about nuclear winter) *and* was associated with a recurring phenomenon (periodicity) whose implications potentially altered our view of the nature of evolution. The death of the dinosaurs may have served as an object lesson about the possible fate of the human species—a once proud and dominant group brought down instantly in a fiery cataclysm—but the message of the periodicity hypothesis was that such events may be a *regular* feature of the history of life, and that existence on this earth may be a much more tenuous affair than previously suspected.

But it is easiest to let a contemporary observer speak to the cultural significance of the relationship between nuclear anxiety and extinction science at the time. Ellen Goodman, a journalist whose syndicated column was carried during the 1980s by newspapers such as the *Boston Globe* and *Washington Post*, wrote a rather remarkable essay, titled "Musings of a Dinosaur Groupie," that was published on January 3, 1984. In the piece, she described her lifelong fascination with dinosaurs and with theories of their demise, which she evocatively connected to changing cultural perceptions. It is worth quoting from the piece at some length. In characterizing traditional views of the extinction of the dinosaurs, which she herself recalled growing up with, Goodman wrote:

> There was a charming egocentricity to these theories. My dinosaurs were evolution's failure and we were its successes. . . . Evolution drew a reasonable pattern in the universe. Over time, species grew better and better. In the rough justice of nature, the fittest survive.
>
> But the theory didn't survive intact. A few years ago, another generation of scientists offered up evidence about my extinct subjects. The dinosaurs didn't gradually die of their evolutionary flaws. The scientists speculated that 65 million years ago an asteroid struck the earth and produced a worldwide crop failure that did them in. My giant vegetarian, the brontosaurus, was the victim of a climatic disaster, a cosmic accident.
>
> Then, in the past year, two scientists at the University of Chicago reported that such disasters have occurred like cosmic clockwork every 26 million years over the past 250 million years, wiping out huge numbers of life forms. The dinosaurs were just the biggest, most memorable of the victims.

Goodman went on to muse whether "every era gets the dinosaur story it deserves," noting that "scientists are also part of their culture, their times," and that this made them at "one moment or another . . . open to a certain line of questioning, a path of inquiry that would have been unlikely earlier on."

> The scientists of the 19th century — a time full of belief in progress — saw evolution as part of the planet's plan of self-improvement. The rugged individualists of that century blamed the victims for their own failure. . . .
>
> The latest theories may reflect our own contemporary world view. Surely we are now more sensitive to cosmic catastrophe, to accident. Surely we are more conscious of the shared fate of the whole species. . . . Most significantly, another group of scientists warns us that a nuclear war between two great powers would bring a universal and wintry death. One hemisphere is no longer immune from the mistakes of the other hemisphere.
>
> In that sense, the latest dinosaur theory fits us uncomfortably well.

"Our" dinosaurs died together in some meteoric winter, the victims of a global catastrophe. As humans, we fear the same fate (Goodman 1984).

One interesting feature of the periodicity hypothesis is that it brought back, in a sense, the narrative of rise, flourishing, and decline present in late-nineteenth- and early-twentieth-century theories of orthogenesis and cyclical history. In contrast to those earlier theories, however, the new understanding of mass extinctions, whether or not they occurred with regular periodicity, emphasized an essential arbitrariness in the history of life. Trilobites and dinosaurs did not "deserve" to die: they were simply caught up in larger forces over which they had no control. No doubt this resonated psychologically with average citizens who felt helpless in the face of impersonal political forces holding the power of life and death at the push of a button; but it also perhaps provided a curious kind of comfort. One message the extinction scientists stressed was that, despite its precarious existence on a tiny rock in an implacably hostile universe, somehow life itself seems to have managed to hang on, and even thrive—at least so far.

If anything, new theories about extinction appeared to repudiate elements of the inherent ruthless competitiveness implicit in the Darwinian account of nature. If extinction is viewed as essentially arbitrary, and not the outcome of a "fair game" in which survival is synonymous with success, it would seem more difficult to celebrate human tendencies toward greed, exploitation, and aggression as being products of a "natural" order. If the 1980s was a decade of great geopolitical anxieties and unprecedented economic disparities, it was also a time when many long-standing assumptions were challenged. Though billed as a triumph of ideology, the Berlin Wall came down in 1989 in part because ordinary citizens simply refused to follow the story their leaders had been acting out for years, and opted instead for community and openness rather than suspicion and division. And as much as the vaunted materialism and acquisitiveness of American culture of the time was celebrated in popular culture, it was as often as not the source of suspicion and criticism—whether in Madonna's 1984 song "Material Girl," or the character Gordon Gekko's famous line "Greed is good," in the 1987 film *Wall Street.*

Media accounts of the extinction debates tended, in fact, to amplify the sense in which new theories challenged this older view of nature. Ellen Goodman's column described contemporary views as overturning a perspective where "in the rough justice of nature, the fittest survived"; it reminded readers that "the astronauts travel into space and report back that they see no national borders." Other popular descriptions highlighted extinction theories as being a direct challenge to Darwinism itself. One 1984 *Washington Post* article announced, "The new emphasis on extinction stands in contrast to Darwin's proposition that evolution was a response to competition for scarce resources"; a 1985 *Time* magazine report suggested that the cyclical extinction hypothesis "call[s] into question the current concept of natural selection" (Rensberger 1985).

These media accounts may have somewhat distorted the scientific message, but some paleontologists were attentive to the ways that mass extinction theory upset previous assumptions. In a discussion titled "Some Implications of Mass Extinction for the Evolution of Complex Life," for example, Sepkoski drew attention to the constructive role that mass extinction has played, noting that "it may prove that total stability is actually detrimental to the evolution of complex life," since "perturbations of the biotic environment . . . may actually be essential to ensure the continuation of evolutionary experiment" (Sepkoski 1985, 230). And in a popular essay in *Discover* magazine in May 1984, Gould wrote that "it makes little sense, though it may fuel our desire to see mammals as inevitable inheritors of the earth, to guess that dinosaurs died because small mammals ate their eggs" (Gould 1984a, 68). On the other hand, Gould noted that the close association between dinosaur impact hypotheses and nuclear Armageddon could have a salutary effect on geopolitical tensions: "I am heartened by a final link across disciplines and deep concerns. . . . A recognition of the very phenomena that made our evolution possible by exterminating the previously dominant dinosaurs and clearing the way for the evolution of the large mammals, including us, might actually help save us from joining those magnificent beasts in contorted poses among the strata of the earth" (Gould 1984a).

Media interest in extinction was also heightened by a spectacular new hypothesis that was emerging in early 1984 to explain Raup and

Sepkoski's proposed extinction periodicity: cyclical comet showers striking the solar system every twenty-six million years, triggered by some undiscovered extraterrestrial phenomenon. This idea was actually proposed independently by two groups of astronomers; remarkably, both papers were published in the same issue of the journal *Nature* in January of that year. The first paper, by Michael Rampino, Richard Stothers, and other colleagues, speculated that a transit of the solar system vertically through the plane of the Milky Way galaxy might bring our local neighborhood into periodic contact—roughly every thirty million years or so—with gas and dust that could disturb the Oort comet cloud, a hypothetical disk containing billions or perhaps trillions of planetesimal bodies located far beyond the furthest edge of the solar system. This might produce periods lasting up to a million years during which the risk of impact on earth would be dramatically heightened, potentially explaining the regular periodicity of extinctions.[26] The second article, which received significantly greater attention—in part because it was received prior to the one by Rampino and Stothers and was thus awarded priority—was authored by the astronomers Marc Davis, Piet Hut, and Richard Muller. It proposed essentially the same effect as Rampino et al., with an alternative mechanism that was even more speculative: a hypothetical red or brown dwarf star orbiting the solar system on an eccentric orbit that passed through the Oort cloud every twenty-six million years.

The media was instantly taken with the notion of this "dark companion to the sun," which its authors colorfully named Nemesis, "after the Greek goddess who relentlessly punishes the excessively rich, proud, and powerful" (Davis, Hut, and Muller 1984, 715). Sometimes referred to in the popular press as a "death star" (an obvious reference to the *Star Wars* trilogy popular at the time), the Nemesis hypothesis injected a sense of menace and inescapable doom to discussions about extinction.[27] As the science writer Denis Overbye put it in a May 1984 article in *Discover*, "Ever since human beings looked to the skies, comets, with their long glowing tails blazing through the night, have portended doom. Now it seems that these primordial fears may have a basis in reality" (Overbye 1984, 26). Many scientists and some media outlets regarded the theory with skepticism or even scorn; a notable *New York*

Times editorial remarked that "astronomers should leave to astrologers the task of seeking the causes of earthly events in the stars." But the public was fascinated by the idea, even though reporters hastened to reassure their readers that the next return of Nemesis was nearly thirteen million years away.[28]

Gould, who frequently championed both the Alvarez scenario and the periodicity hypothesis in both popular and scientific articles, had a somewhat different take. While he supported investigating the basic phenomenon—and did not find a hypothetical companion star necessarily implausible—he made a plea to Davis, Hut, and Muller: "If Thalia, the goddess of good cheer, smiles upon you and you find the sun's companion star, please do not name it (as you plan) for her colleague Nemesis, [since] she represents everything our new view of mass extinction is struggling to replace—predictable, deterministic causes afflicting those who deserve it" (Gould1984b, 18–19). Gould reasoned that "if mass extinctions are so frequent, so profound in their effects, and caused fundamentally by an extraterrestrial agency so catastrophic in impact and so utterly beyond the power of organisms to anticipate," then scientists must develop "new and undiscovered rules for perturbations" rather than "laws that regulate competition during normal times." As an alternative to "Nemesis," Gould proposed the name Siva, "the Hindu god of destruction, [who] forms an indissoluble triad with Brahma, the creator, and Vishnu, the preserver," and who, unlike Nemesis, "does not attack specific targets for cause or for punishment." In Gould's reasoning, Siva better personified the sense that "mass extinctions are not unswervingly destructive in the history of life," but also are a "source of creation as well" by providing "the primary and indispensable seed of major changes in life's history." Thus, unlike Robert Oppenheimer's invocation of Siva as simply "destroyer of worlds" (a reference Gould knew very well), Gould's proposal reflected the sense in which the new view of extinction acknowledged that "destruction and creation are locked together in a dialectic of interaction."

In this sense, by the mid-1980s, both the science and the culture of extinction had found a new context for anxiety and a new sense of moral lesson. The regular occurrence of mass extinctions in the history of life did indeed suggest that, as Raup put it, "our planet may not be such a safe place"—but it also led to a growing awareness that, if extinction is effectively *arbitrary*, at least with respect to the selective conditions of the environment prior to the extinction event, then it is rarely ever *necessary* (Raup 1991, 5). The pessimistic message this presented was that no species is too dominant, too well adapted, or too widespread to be safe in the event of the catastrophe. If it happened to the dinosaurs, it could happen to us. And the particular scenario proposed for the extinction of the dinosaurs did sound uncomfortably similar to descriptions of nuclear Armageddon. As Walter Alvarez later put it in his popular account of his hypothesis, the asteroid arrived with the energy "of a hundred million hydrogen bombs," creating scenes like those vividly portrayed in *The Day After*. Alvarez went on to describe the imagined scenario:

> In the zone where the bedrock was melted or vaporized, no living thing could have survived. Even out to a few hundred kilometers from ground zero, the destruction of life must have been nearly total. . . . Animals living just over the horizon first witnessed a flash of light in the sky, then a last moment of calm. Then, as the ground began to shake uncontrollably from the passing seismic waves, the sky itself turned lethal. . . . Soon the Earth's surface itself became an enormous broiler—cooking, charring, igniting, immolating all trees and all animals which were not sheltered under rocks or in holes. . . . Entire forests were ignited, and continent-sized wildfires swept across the lands. The ejecta particles had barely fallen to Earth and the lethal, incandescent sky returned to normal, when the air was blackened by rising plumes of soot from fires which were consuming the forests and removing the oxygen from the atmosphere (Alvarez 1997, 11–12).

As terrifying as this vision was, however, it could also act as inspiration to avoid the dinosaurs' fate, as Gould and others pointed out. The dinosaurs could not escape their asteroid, but humans might yet take action to stave off their own extinction. The pessimistic reading of mass extinction theory thus also offered a more optimistic corollary: While the flourishing of mammals and eventual rise of human civilization may have been all just the result of a "cosmic accident," there was no reason to suppose that any species, including our own, was foreordained to die. In an indirect way, extinction theory, combined with nuclear winter projections and growing public awareness of the magnitude of the catastrophe that would result from even a limited nuclear exchange, certainly had an influence on easing geopolitical tensions and encouraging nonproliferation and disarmament in the late 1980s and the 1990s. More directly, it encouraged awareness and action around other potential crises—such as anthropogenic climate change and biodiversity depletion—that began to replace nuclear war as major cultural and political anxieties. As David Jablonski put it, "The mass extinctions in the fossil record have compelling implications for the plight of today's wildlife and for the survival of the human species"—namely, "that major upheavals can and do occur and that such biological crises can be rapid, irreversible, and unpredictable." Warning that humans were "on the brink of causing, single-handedly, the worst mass extinction in 65 million years," Jablonski urged, "It is up to us, as beneficiaries of the last major mass extinction, to reverse this trend . . . before many of the species we hold dear—including our own—go the way of the dinosaur" (Jablonski 1986c, 61–63). This was the beginning of a new extinction imaginary—discussed in depth in the next chapter—that transferred the anxiety about catastrophic human activities to an overt call for action and activism that has characterized a new extinction discourse, and which persists to this day.

Extinction, History, and Culture

We have so far dealt with the direct relationship between the science of mass extinction and the culture and politics of the late Cold War era as a fairly overt sharing of imagery, rhetoric, and even empirical evidence about the consequences of major catastrophic events in the physical and biological environment. But there are also other ways in which the late 1970s and 1980s were a time of cultural confrontation with extinction and catastrophism in less literal, though nonetheless important, forms of expression. One sense in which this manifested was in the notion that late modernity is an intrinsically "catastrophic society" in which threat and risk have been internalized in political beliefs, psychological reactions, and literary and artistic expressions to the extent that structures of meaning and certainty have broken down. This was to some extent a continuation of the gloomy pessimism of Modernism discussed in chapter 3 of this book. But insofar as it took shape in an era of art and philosophy that had also transcended the early twentieth century's nostalgia for those lost structures—the "postmodern" era—these new views had a distinctive character that in many ways complemented developments in the science of extinction.

Postmodernity itself—or postmodernism—is hard to define, and indeed perhaps intentionally resists definition. But, as introduced by continental European thinkers like Jean-François Lyotard, Jacques Derrida, Michel Foucault, and others (not all of whom would have consented to be grouped under this label), it emphasized an essential ambivalence around extracting stable categories or "meaning" from texts or discourse, owing to the inherent instability of language. As Lyotard, for example, famously declared in his seminal 1979 book *The Postmodern Condition*, "I define *postmodern* as incredulity toward meta-narratives" (Lyotard 1979, xxiv). These included the supposed certainties of science as well as the knowledge structures of politics, philosophy, art, and other forms of cultural discourse. As applied to the study of texts— indeed, one feature of postmodernism was to expand the notion of "text" to encompass virtually any form of human expression—this skepticism was often expressed by Derrida, Jean Baudrillard, and others as a rejection of the notion that meaning is grounded in some objective external reality. Since language is understood to be constitutive of our perceived reality, and since words are seen merely as "signs" with no stable relationship to objective referents, then what we experience is a simulation—or, as Baudrillard put it, a "simulacrum"—of meaning in which images are reproduced and recycled without retaining any reference to some original.

This notion is admittedly quite abstract, and while it was popular among students and intellectuals especially during the 1980s and early 1990s, it should not be overstated as a broad cultural phenomenon. Postmodernist philosophy did engage directly with some of the central themes of contemporary extinction imaginary, however, and it gives an interesting perspective on the wider cultural manifestation of the issues in this chapter. If a major characteristic of Modernism was the notion that the values and societies of the West were in a state of decline and disintegration, then postmodernism often adopted the perspective that those very structures that had formerly offered meaning—in art, politics, philosophy, and even science—were fractured beyond repair. Modernism often presented society as waiting in anticipation of some apocalypse ("And what rough beast, its hour come round at last, / Slouches towards Bethlehem to be born?"); postmodernism, in contrast, saw the apocalypse as having already happened. The perceived cataclysm was understood to some extent as a metaphorical notion, but it also drew inspiration and imagery from tangible twentieth-century political and environmental events. As Simon Malpas puts it in a study of postmodernism, "The threat of the obliteration of all existence, whether brought about by nuclear war or natural catastrophe, has weighed on ideas of what it is to be part of a community or society, and even what it is to be human, forcing thoroughgoing reconceptualisations of some of the most basic categories of philosophical, social and political thought" (Malpas 2005, 34).

Implicit in many philosophies of history associated with postmodernism, including those of Jürgen Habermas and Foucault, is a rejection of the notion that history proceeds toward ever better models of rationality and social arrangement. The sense of historical continuity is eroded with the departure of guiding metanarratives, which are often considered a product of an Enlightenment transfer of Christian providential theology to a secularized view of human progress (particularly embodied in the nineteenth-century German philosopher G. F. Hegel's progression of history toward an "absolute"). In this sense, it is fairly easy to draw parallels between post-Enlightenment historiographies of human and natural history; just as Hegel or August Comte viewed human history as a linear progression toward greater rationality and self-awareness, Darwinian evolutionary history—if not exactly reflecting Darwin's own view, particularly in the interpretation of Herbert Spencer—saw the progression of life as a march towards greater complexity and order.

From this perspective, the postmodernist critique aligns comfortably with contemporary reinterpretations of the history of life. In the context of the new understanding of mass extinctions, life's history becomes less a steady, continuous stream than a series of distinct episodes, broken by drastic upheavals that reset environmental and biotic conditions. The notion of "punctuation" or "rupture" as a feature of historical development had fairly wide currency in both biology and historiography in the 1970s and 1980s. In 1972, for example, Stephen Jay Gould and Niles Eldredge proposed a controversial model they labeled "punctuated equilibria," in which evolutionary lineages were characterized as being mostly unchanging and static except for relatively brief moments of rapid change when new species were produced in a geological instant.[29] In similar fashion, the French theorist Michel Foucault had argued in his 1966 book *The Order of Things* (published in English translation in 1970) that human history resolves to a series of distinct "epistemes," or worldviews, punctuated by ruptures that have altered basic notions of truth and representation (Foucault 1966). Foucault's interpretation of history shares some marked similarities with the philosopher Thomas Kuhn's view of science as presented in his 1962 *Structure of Scientific Revolutions*, which argued that the history of science is composed of a series of distinct "paradigms," which in more radical interpretations (for example, by the philosopher Paul Feyerabend) have altered conditions for truth and meaning.[30] Notably, Foucault flirted with a kind of metaphorical notion of extinction, concluding at the end of *The Order of Things* that "the figure of man" is a fairly recent Enlightenment concept which, should the conditions that brought it into being erode, "would be erased, like a figure drawn in sand at the edge of the sea" (Foucault 1966, 386–87). While it certainly would be possible to make too much of the similarities between these scientific and philosophical reinterpretations of historical change, Gould himself (who frequently invoked philosophers when presenting paleontological ideas) commented on their similarities, picking out Kuhn and Foucault in particular. In an essay titled "Toward the Vindication of Punctuational Change," in which he broadly surveyed challenges to geological uniformitarianism (and name-checked Foucault and Kuhn), Gould concluded that while he did not "know how much of this new fascination for punctuational change resides in the stresses of our general culture, . . . our uncertain world of nuclear armaments and deteriorating environment must also encourage a departure from gradualism" (Gould 1984c, 31).

In a variety of ways, 1980s-era critiques of narratives of progress resulted in a discourse around the notion of "the end of history." This did not necessarily mean the literal end of human civilization—through extinction, for example—as much as an end, as Malpas puts it, to our "ability to form a narrative from [events in the past] that demonstrates their coherent, developmental logic and points to a utopian future in which the conflicts and contradictions between them will have been re-solved" (Malpas 2005, 89–90). This notion could manifest in a variety of philosophical viewpoints, not all of which could be described as "post-modern." The neoconservative theorist Francis Fukuyama, for example, argued in a much-discussed 1989 essay titled "The End of History" (ex-panded to a book in 1992) that the fall of the Berlin Wall represented a kind of culmination of Western democratic ideals, which he described as "the end point of mankind's ideological evolution and the universal-ization of Western liberal democracy as the final form of human gov-ernment" (Fukuyama 1989, 4). While he certainly did not see this as an entirely unwelcome development, he also argued that "the end of his-tory will be a very sad time," since the disappearance of ideology would mean that "daring, courage, imagination, and idealism, will be replaced by economic calculation, the endless solving of technical problems, en-vironmental concerns, and the satisfaction of sophisticated consumer demands." Strikingly, he concluded that "in the post-historical period there will be neither art nor philosophy, just the perpetual caretaking of the museum of human conflict between states" (Fukuyama 1989, 18).

Fukuyama was no postmodernist, and was indeed roundly criticized by Derrida and others for what they perceived as Western triumphal-ism. But in a sense his vision of the end of history resonates with the postmodernist argument that we have reached a stage where we are simply rearranging images of the past in a kind of pastiche without de-veloping any new structures of meaning. This was a central argument of Baudrillard's 1981 study of popular culture, *Simulacra and Simula-tion*, where he argued that mass communication has dissolved all stable notion of reference into an endlessly self-referential "hyperreality." In a short essay published in 1989 titled "The Anorexic Ruins," Baudrillard compared the perpetual recycling of hyperreality to "cancerous metastases," and described the "merciless short circuit" as "a catastrophe in slow motion."[31] But unlike previous cultures, Baudrillard argued, our own has lost even the possibility of some kind of final "reckoning, denouement, and apocalypse"—some possibility of either destruction or rebirth—since "we have already passed it unawares and now find ourselves in the situation of having exhausted our own finalities" (Baudrillard 1989, 33–34). Here Baudrillard invoked the nuclear anxieties of his age with a striking statement: "Everything has already become nuclear, faraway, vaporized. . . . The explosion has already occurred, the bomb is only a metaphor now." In this view, the true catastrophe would not be the end of our existence—after all, we would not be around to experience it—but our continued existence in an "amnesiac world" capable only of recycling images of its own past. The film *The Day After*, he argued, did not conjure up the horror of a possible fate; rather, he claimed that "this film itself *is* our catastrophe," since "it says that the catastrophe is already there, that it has already occurred *because the very idea of the catastrophe is impossible*" (Baudrillard 1989, 37). What he appears to mean by this is not that nuclear weapons do not exist or that nuclear war is impossible, but that our society has lived in a state of perpetual catastrophe for such a long time that we now exist in "a perpetual simulation of crisis" without having developed a new philosophy or means of expression to move beyond it (Baudrillard 1989, 42).

If Baudrillard's analysis recasts catastrophe as a metaphorical concept, his formulation nonetheless touches on a theme present in other, more tangible assessments of 1980s political culture. One example is the concept of "risk society" developed by the German sociologist Ulrich Beck, a prominent public intellectual whose 1986 book of the same name (its English translation appeared in 1992) argued that modernization has inherently led to "the social production of *risks*" as consequence of the generation of wealth (Beck 1986, 19). Beck described the current political climate as "reality that is out of joint," destabilized by the proliferation of human-engineered "destructive forces" that "endanger *all* forms of life on this planet" and are able to "outlast generations" (Beck 1986, 10, 22). While Beck was much more concerned than Baudrillard with tangible manifestations of catastrophe—ecological disaster or industrial accident—like Baudrillard, he highlighted the social effects of existing in a state of perpetual crisis, where "the *state of emergency* threatens to *become the normal state*," and where the possibility of transformative crisis is foreclosed: "The risk society is thus not a revolutionary society, but more than that, a *catastrophic society*" (Beck 1986, 78–79).

The diagnosis that Beck, Baudrillard, and others provided was thus as much about the spirit or psychology of catastrophic or postapocalyptic society as it was about the danger of actual, immediate physical cataclysm. This signals an important turning point in the history of catastrophic thinking: while threats like nuclear war or environmental disaster continued to have a prominent role in the popular imagination, the sense of the time scale on which they were anticipated or experienced began to be expanded, and their harmful effects were projected onto the present as well as onto an imagined future. As we will see in the next chapter, this became a central theme in extinction discourse from the mid-1980s onward, particularly in discussions of biodiversity loss and anthropogenic climate change. But it was also manifested in other cultural forms including, for example, the dramatic growth in the popularity of postapocalyptic science fiction during the late 1970s and the 1980s, with stories that increasingly focused on characters attempting to cope with life in catastrophic landscapes, rather than with catastrophic events as the culmination of a narrative.

Even fictionalizations of nuclear war began to reflect this shift. Whereas *The Day After* ended in the immediate aftermath of a nuclear exchange, the 1984 BBC film *Threads* (often regarded as much superior to *The Day After*) followed its central characters, a young woman named Ruth and her infant daughter, through a series of vignettes set days, weeks, months, and ultimately years after the war. In addition to offering the first cinematic representation of nuclear winter, *Threads* presented a decidedly ambivalent vision of the survival of humanity. The scenes set weeks or months after the initial explosions followed the characters through a desolate wasteland accompanied by titles accounting the numbers of the dead, but in the film, humanity does not immediately die out. Rather, we are forced to contemplate the awful circumstances

of Ruth's survival, in which she scrounges for food and shelter, bart-
ers sex for dead rats to eat, and develops symptoms of radiation sickness. Viewers are informed that, three to eight years after the attack, the
population has reached its minimum, thus implying that final extinction
has been staved off. But the society Ruth and her daughter occupy is dismal: existence is barely at a subsistence level, people are left to scavenge
clothes and other necessities from ruined cities, and children speak a
kind of grunting language and scuffle for food like animals. When, ten
years on, Ruth finally succumbs to her illness, her daughter emotion-
lessly removes her few valuable items and carries on.

This new perspective was found in other, more commercial depictions as well. A prime and extremely influential example is the 1979
Australian cult favorite *Mad Max*, which tells the story of a policeman
seeking vengeance against a group of motorcycle-riding thugs who have
killed his family, and which is set "a few years from now," in what appears to be some kind of postapocalyptic wasteland. The context of the
dystopian setting is never explained, however, and the film's climax is
Mel Gibson's character, Max, defeating the gang leader, after which he
simply drives off into the distance. James McCausland, who cowrote
the film, later recalled being inspired not by nuclear apocalypse but by
the impact of the 1970s oil crisis on Australian society, and basing the
grim scenario on "the assumption that nations would not consider the
huge costs of providing infrastructure for alternative energy until it was
too late" (McCausland 2006). Although sequels like *The Road Warrior*
(1981) and *Mad Max: Beyond Thunderdome* (1985) would later flesh
out some details—hinting, in the third installment, at a nuclear war—
the fictional setting in which the story takes place is self-contained
and static; its characters have accepted the bleak, terrifying scenario
in which they live, and the films are about their struggles, largely free
from nostalgia for the lost world or hope of redemption for a new one.

In many ways, the new spirit of anxiety that developed during the
1980s was captured in a series of long essays published by Jonathan
Schell across successive issues of the *New Yorker* in February of 1982,
and published later that year as a book titled *The Fate of The Earth*.
Schell, a longtime staff writer for the magazine, had an established
reputation as a political reporter and critic, having covered the Viet-

nam War and the Watergate scandal, and was described by the environmental activist and scholar Bill McKibben after his death in 2014 as having been "for many years a central figure both at this magazine and in the intellectual life of the nation" (McKibben 2014). But Schell's overwhelming concern—an obsession, even—was in warning the public about the dangers of nuclear weapons before it was too late. The essays collected in *The Fate of the Earth* registered as some of the most powerful and resonant arguments yet made; the *New York Times* review of the book called it "a work of enormous force" and "an event of profound historical moment" (Erikson 1982).

The book is broken into three parts. The first, "A Republic of Insects and Grass," vividly describes the world in the aftermath of a nuclear exchange, emphasizing not just the toll on human populations but the enormous environmental catastrophe that would ensue. The third, "The Choice," outlines the role of the politics of national sovereignty in the deterrence strategy of mutually assured destruction, arguing that the only way out of the standoff is for humans to identify as a collective species rather than as nations and factions. Neither essay presents a particularly original viewpoint, though Schell's accomplished literary style and the urgency of his prose probably accounts for the attention they received. There is nothing in "A Republic of Insects and Grass," for instance, that could not be gleaned from more technical reports, and even the effusive *New York Times* reviewer acknowledged that the argument of "The Choice" had "been said so often before that the sheer mention of it simply sounds naïve."

It is the second essay, titled "The Second Death," that stands out. An exploration of the metaphysical, rather than physical, consequences of nuclear war and extinction, this essay essentially argues that extinction—taken by Schell to be the likely outcome of a nuclear war—would produce two kinds of "death": both "the untimely death of everyone in the world," which "would in itself constitute and unimaginably huge loss," and "a separate, distinct loss that would be in a sense even huger—the cancellation of all future generations of human beings" (Schell 1982, 59). The essay thus departs from other similar discussions, such as Karl Jaspers's *The Future of Mankind*, not only in contemplating the possibility of human extinction and its ethical and political consequences, but in delving into the philosophical and psychological consequences of this awareness: it is an investigation into the *meaning* of extinction. Schell noted that mass extinctions have been a feature of life's past, and that the extinction of humans "would constitute an evolutionary setback of possibly limited extent . . . perhaps no greater than any of several evolutionary setbacks, such as the extinction of the dinosaurs." However, no other species, he assumed, has ever had the ability to contemplate its own extinction; we are unique, having "eaten more deeply of the fruit of the tree of knowledge," and have "caused a basic change in the circumstances in which life has been given to us, which is to say that we have altered the human condition."

One might plausibly argue, as Schell acknowledged, that this "second death" of extinction is "merely redundant," since once our species is extinct there will be nobody to mourn it (Schell 1982, 60). Indeed, he granted that "we, the living, will not suffer it; we will be dead." However, he also noted an apparent paradox in extinction: while it might appear to be "the largest misfortune that mankind could ever suffer," since by definition nobody would be left to experience it, "it doesn't seem to happen *to* anybody, and one is left wondering where its impact is to be registered, and by whom" (Schell 1982, 74). The answer, of course, is that it is the *living* who suffer. Here Schell quoted Montaigne, who wrote: "You are in death while you are in life, for you are after death when you are no longer in life. Or, if you prefer it this way, you are dead after life, but during life you are dying; and death affects the dying much more roughly than the dead, and more keenly and essentially" (Montaigne, in Schell 1982).[32] "We are similarly," Schell argued,

> "in extinction" while we are in life, and are after extinction when we are extinct. Extinction, too, thus affects the living "more roughly" and "more keenly and essentially" than it does the nonliving, who in this case are not the dead but the unborn. Like death, extinction is felt not when it has arrived but beforehand, as a deep shadow cast back across the whole of life. . . . We the living experience it, now and in all the moments of our lives. Hence, while it is in one sense true that extinction lies outside human life and never happens to anybody, in another sense extinction saturates our existence and never stops happening (Schell 1982, 78).

For Schell, extinction is thus "more terrible—is the more radical nothingness—because extinction ends death just as surely as it ends birth and life" (Schell 1982, 63). It is both "the death of death" and "the murder of the future"—and its consequences can only be felt by the living (Schell 1982, 100). It is this condition—the recognition of being "in extinction"—that characterizes the transformation of the extinction imaginary during the 1980s. It was brought about only when people fully absorbed the potential for human extinction (via nuclear war and especially nuclear winter) as well as its environmental and evolutionary consequences, through the new understanding of mass extinction. Extinction is now a "specter" that "hovers over our world and shapes our lives with its invisible but terrible pressure," accompanying us "from birth to death" (Schell 1982, 101). In this sense, as Baudrillard would later put it, "the explosion has already occurred"—or at least it may as well have occurred, since we the living are the ones fated to experience the horror of extinction. The "postmodern condition" is thus aptly described as being "postapocalyptic"; as Schell concludes, "It is the truth about the way we now live."

Conclusion

While Schell's message was potentially quite gloomy, pessimism is not the central message with which I want to conclude this chapter. The larger importance of Schell's diagnosis—that we are now "in extinction"—is what it signifies about the significant transformation of extinction discourse during the 1980s, a shift manifested in scientific and popular imaginations of the causes and consequences of extinction, which altered perceptions of the nature of the threat, the time scale on which it played out, and the role of human agency in its prevention.

In the first place, the science of mass extinction contributed directly to the acceptance of catastrophic change as a regular feature of earth's history. While sometimes referred to as "catastrophism," this new understanding was sometimes also characterized as a "new uniformitarianism," as it was in the title of a collection of essays on sudden geological change published in 1984 (Van Couvering et al. 1984). This emphasized the way in which sudden perturbation was being incorporated into a model of historical change that nonetheless also exhibited signifi- cant regularity and predictability. From the perspective of geology, this new view combined elements of both the Lyellian steady-state and the Cuverian revolution. As Sepkoski's perturbed logistic model showed, the general tendency of the earth's biota is toward a stable equilibrium, but that equilibrium can be and has been disturbed by major crises that have reset ecological and environmental conditions, and which have had major consequences for diversification.

The resulting picture is one of contrasting patterns on different levels of historical scale. Viewed from the perspective of hundreds of millions of years, life on earth actually appears to be remarkably stable, having withstood crises (such as the late Permian event) that destroyed nearly all living species without suffering an absolute decline in diver- sity. At the same time, as Sepkoski's colleague David Jablonski showed, fundamentally different rules may apply during periods of mass extinc- tion, upsetting the Darwinian assumption that, as Gould put it, "order rules as the predictable struggle of individuals translates to patterns of increasing complexity and diversity" (Gould 1984b, 17). In fact, as Jablonski argued in an influential 1986 article, the history of life dem- onstrates two distinct "macroevolutionary regimes": one that applies during normal "background" times, and the other at moments of en- vironmental crisis. Jablonski emphasized that "mass extinctions are not simply intensifications of processes operating during background times," but are processes "qualitatively as well as quantitatively different in their effects," and that ultimately are responsible for "shap[ing] large- scale evolutionary patterns in the history of life" (Jablonski 1986a, 129).

A number of important implications followed from this new under- standing of mass extinction. On the one hand, as Raup put it, "Our planet might not be such a safe place." While potentially unsettling, this message was not news to the generations who had lived through two world wars, genocides, environmental catastrophes, and social up- heaval, and who had grown up in the shadow of the bomb. If Cuvier's catastrophism was ultimately rejected by his nineteenth-century con- temporaries because it was inimical to Victorian notions of stability and progress, then clearly by the late twentieth century cultural assumptions had become much more receptive to an inherent sense of instability and catastrophe. The work of Raup and Jablonski also highlighted the sense in which past success could not necessarily guarantee future survival, since "many traits of individuals and species that had enhanced survival . . . during background times become ineffective during mass extinctions." This fact undercut assumptions about inherent directional progress, and also stressed the essential unpredictability—or contingency—of the pattern of life's history: as a result of mass extinctions, "evolution is channeled in directions that could not have been predicted on the basis of patterns that prevailed during background times" (Jablonski 1986a, 132).

On the other hand, the paleontologists also emphasized that major upheavals had potential benefits, at least from the perspective of the overall diversity of life. As Sepkoski argued, "In the absence of mass extinction . . . macroevolution would be confined to the slow process of anagenesis [species evolution without branching] and evolutionary novelties would appear rarely at best. . . . Only mass extinction would break this stagnation by clearing ecospace for the radiation of new lineages." Sepkoski was implying that without mass extinction, life might not be very diverse or complex (Sepkoski 1985, 230). Whether or not this was perceived overall as positive or negative is, then, a matter of perspective. The Permian extinction was bad news for the trilobites but good news for clams; the Cretaceous-Tertiary event was bad for the dinosaurs but good for mammals. The story, though, does not yield a moral about winners and losers as easily as does the traditional Darwinian account; the trilobites and dinosaurs did not "deserve" to become extinct, nor did clams and mammals deserve to survive. While each group had genetic traits that contributed to ultimate survival or failure, none could have anticipated the selective conditions that were suddenly applied when a mass extinction struck. As Raup quipped, it was simply "bad luck to have bad genes."

Another major feature of the emerging scientific understanding of extinction was an increasing focus on the relationship between ecological diversity and stability. While Sepkoski's long-term analysis suggested that stability was perhaps "detrimental" to evolutionary experiment and diversification, it became increasingly clear just how important stability was to the maintenance of existing levels of diversity. During times of mass extinction, levels of standing diversity plummeted, in part because complex relationships of interdependency within ecological systems were disturbed. Again, the value attached to this phenomenon is a matter of perspective: while from a long-term evolutionary vantage the periodic collapse of diversity may have opened up opportunities for "experiment," from the perspective of the inhabitants of the affected ecosystems, these events were catastrophic.

It was not lost on the scientists who contributed to this new view that lessons could be drawn from the past for our human present. Jablonski observed in a 1986 essay,

> The mass extinctions in the fossil record have compelling implications for the plight of today's wildlife and for the survival of the human species. The fossil record is telling us that major upheavals can and do occur and that such biological crises can be rapid, irreversible, and unpredictable. Once a species is extinct or a network of interacting species falls apart, it is gone forever (Jablonski 1986b, 61).

As early as 1983, at the meeting on the "Dynamics of Extinction" in Flagstaff, this message was adapted directly to the present-day environmental crisis. In a paper titled "What Is Happening Now and What Needs to Be Done," Ehrlich argued, "The earth's biota now appears to be entering an era of extinctions that may rival or surpass in scale that which occurred at the end of the Cretaceous.... For the first time in geologic history, a major extinction episode will be entrained by a global overshoot of carrying capacity by a single species—*Homo sapiens*" (Ehrlich 1986, 158). At the same conference, the ecologist Daniel Simberloff addressed the crisis of deforestation in tropical rain forests, and concurred with Ehrlich that "the imminent catastrophe in tropical forests *is* commensurable with all the great mass extinctions except for that at the end of the Permian" (Simberloff 1987, 177–78). Nor were the ecologists alone in making such claims; Jablonski as well had warned, bluntly, "Our species ... is on the brink of causing, single-handedly, the worst mass extinction in 65 million years" (Jablonski 1986b, 63). In identifying current ecological crises with past mass extinctions, Ehrlich, Simberloff, and their paleontologist colleagues were thus gesturing toward a new framework for understanding the impact of human beings on their environment: humans could now be understood as agents of global environmental change—perhaps on a par with the geological or extraterrestrial forces that have caused the pass extinctions of the past—as well as its potential victims. In the iconography of the emerging extinction imaginary, we are both the asteroid and the dinosaur.

A central outcome of this rhetorical turn was that diversity itself became identified as the entity threatened by mass extinction. Ehrlich, for example, did not single out one species or another for special concern, but rather argued that "the very future of humanity depends on preserving organic diversity" as a whole, in part because of "the utter dependence of our species on the free services provided by ecosystems" (Ehrlich 1986, 162, 157). The notion that biological diversity is an inherent source of health and stability for ecosystems will be explored in much more detail in the next chapter, but a vitally important point to emphasize here is the close dependence that the emergence of "biodiversity" as an "inherent value" in the language of conservation biology and politics had on the developing science of mass extinction by paleontologists. It was paleontologists like Raup, Sepkoski, and Jablonski who had redefined the study of mass extinction as a study of patterns in taxonomic diversification, and who likewise had explored the ecological and evolutionary consequences of mass extinctions from which ecologists and biologists drew. In a somewhat later study, Jablonski, for example, observed that paleontology is "our only direct source of information on how biological systems respond to large-scale perturbations and thus can provide important insights into potential outcomes if habitat destruction or climate change proceeds unchecked." One of his most significant findings was that mass extinctions tended to favor "weedy species . . . rats, ragweed, and cockroaches," capable of surviving in a variety of environmental conditions, at the expense of "the larger number of species that are more useful to humans as food, medicines, and genetic resources" (Jablonski 1991, 755).

Two final points can be made about the preceding discussion. The first is that, in the evolving conversation about the modern-day extinction crisis in biological diversity, the term "resource" emerged as a multivalent concept. Individual species of plants and animals can be identified as resources—as they were for many years in earlier conservation discourse—because of their utilitarian or aesthetic value to human beings. But biological diversity only came to be seen as a resource in itself when the ecological viewpoint developed during the 1950s and 1960s (discussed in chapter 4) that identified the stability of ecosystems with the diversity of their inhabitants was projected onto an understanding of global and historical patterns of diversification and extinction in the 1970s and 1980s. In this perspective, the diversity of life is not only a cornucopia of useful materials from which humans can draw, but also a vital hedge against unpredictability and environmental collapse. A mass extinction is understood, by definition, as a cascading phenomenon that takes place when any portion of the foundation on which ecosystems are stabilized is removed; it is not defined by how important any individual group that dies may seem to us.

Second, in this perspective, biological diversity, like genetic diversity, is understood to be a reserve not just of things but of "information" or "potential." In *The Fate of the Earth*, Schell noted that if we can understand the life of an individual creature to be "information, and death is the loss of information," then in the extinction of a species "the sources of all future creatures of those kinds are closed down, and a portion of the diversity and strength of terrestrial life in its entirety vanishes forever" (Schell 1982, 56). As Sepkoski found in his study of Phanerozoic diversity patterns, when life rebounds following extinction events, it tends to diversify within a narrower range of possible forms. After all, since evolution does not repeat itself, the removal of a higher taxon means removing all of the genetic information contained within its individual lineages, leaving less raw material to work with. And what tend to remain are what Jablonski calls "weedy species" which, like many of the animals and plants transplanted by Europeans into their colonial possessions, can dominate large environments to the exclusion (and extinction) of more varied, specialized forms of life. Mass extinction, then, is not just the temporary reduction of life's variety, but the potentially permanent depauperization of the earth's biota.

This, then, is the context for the final chapter in our story: an extinction imaginary combining a new view of the causes and consequences of extinction and a new sense of the role human beings play in the maintenance of diversity and stability in our world. It is also the envisioning of a new slow-motion catastrophe whose effects have already begun to be felt, but whose ultimate consequences may not be known for many years or decades. It is the foundation for the discourses of the "Sixth Extinction" and the "Anthropocene"; but, more broadly, for a new perspective on the place of humans in their natural world—one in which our sense of intrinsic importance to this planet is challenged at the same time as the impact of our agency is magnified.

6

A SIXTH EXTINCTION? THE MAKING
OF A BIODIVERSITY CRISIS

The recognition, just in recent years, that mass extinctions do
not represent the processes of background extinctions writ
large must rank as one of the most important discoveries in
evolutionary biology of this century. Whatever their cause, mass
extinctions operate by different rules from those prevailing
during background extinction. Darwinian evolution, important
in background times, is suspended during biotic crises.

—Richard Leakey and Roger Lewin, *The Sixth Extinction* (1995), 228

Humanity has initiated the sixth great extinction, rushing to eternity
a large fraction of our fellow species in a single generation.

—E. O. Wilson, *The Diversity of Life* (1992), 32

It is rare that the origin of a significant cultural movement can be located
in a single event—history is normally much too messy and complex for
such easy explanations. Indeed, in the case of the movement around
what is now widely understood to be the "biodiversity crisis," this is
very much the case: as this book has argued, the development of late-
twentieth and early-twenty-first-century attitudes and beliefs concerning the value of biological diversity and the threat of anthropogenic extinction have had a long, complicated history stretching back more than
two hundred years. However, one element of this history, the invention

of the term "biodiversity," can be traced to a single point of origin, and it can be argued that the emergence of the term itself and the rapid assimilation of the concept into wide political and cultural currency went hand in hand.

The event in question was the "National Forum on BioDiversity" held in Washington in September 1986, which attracted an audience of several hundred scientists, policy makers, journalists, and members of the public to hear some sixty speakers discuss the causes and consequences of human-caused extinction of plant and animal species. The meeting was cosponsored by the National Academy of Sciences and the Smithsonian Museum, and was the brainchild of National Research Council senior staff officer and plant physiologist Walter G. Rosen, who enlisted E. O. Wilson as the intellectual driving force. As was later reported by both Wilson and Rosen, during the planning stages Rosen was concerned that the phrase "biological diversity," in circulation since about 1980, was too much of a mouthful. In a letter to Wilson he wrote, "We can save three syllables by taking the logical out of biological." Over Wilson's initial objections, the neologism was adopted for the title of the conference.[1] Wilson may not have loved the contraction, but the term stuck—as did the public and scientific concerns raised during the conference—in the eventually published proceedings (which Wilson edited) and in a coordinated campaign of journal and magazine articles, popular books, public lectures, and policy initiatives during the following years.

Viewed from one perspective, biodiversity awareness burst on the scene suddenly and with rapid success. Wilson's paean *The Diversity of Life* was a best seller when released in 1992, and it has remained continuously in print to this day. And at the 1992 "Earth Summit" held in Rio de Janiero, more than 150 nations signed the United Nations "Convention on Biological Diversity," which formally acknowledged biological diversity as a cultural and economic resource.[2] Within a decade of the Washington forum it was broadly accepted that human activities—most prominently tropical deforestation, but also anthropogenic climate change, industrial agriculture, human population explosion, and global development—had precipitated a crisis in which as many as half of all existing species of plants and animals could become extinct within a century. By the mid-1990s, the crisis had acquired another, more foreboding name—the "Sixth Extinction"—that has effectively drawn great public attention to biodiversity loss by connecting the present depletion to the mass extinctions of the geological past.[3] As the literary scholar Ursula Heise has observed, in the new millennium "the threat of mass extinction now often features as one of several global ecological crises, right behind climate change in the urgency of action it requires" (Heise 2016, 21).

At the same time, however, both the biodiversity crisis and the "Sixth Extinction" concept—along with the wider set of global environmental concerns grouped together under the label of the "Anthropocene"—are simply the most recent manifestation of the cultural and scientific discourse around extinction and humankind's future we have been following through this entire book. While the terms, anxieties, and imagined consequences of the current dialogue are novel in many ways, they also show the strong imprint of a set of concepts and concerns that have been in circulation since the 1950s and 1960s if not earlier: fear of a catastrophic end to civilization, awareness of the interconnectedness and fragility of ecosystems, a growing valuation of diversity as a bulwark against unpredictable change, and of course appreciation of the reality of mass extinctions as a major feature in the history of life. As the environmental historian Timothy Farnham aptly put it, "The rise of popularity of the biological diversity cause was not necessarily a paradigm shift, but it was a confluence of values and concern that had been fostered over time, coming together in one concept that represented the protection of the living components of the natural world" (Farnham 2007, 12). It is this confluence of values that forms the center of the current extinction imaginary.

While we might consider the biodiversity crisis and Anthropocene concepts to be the apotheosis of the post–Second World War extinction discourse traced in chapters 4 and 5 of this book, many of its central preoccupations have been reframed and relationships redefined.[4] For one thing, while a considerable anthropocentrism is retained in both concepts (biological diversity is still often defined in terms of its value to humans, and Anthropocene proponents would like to name a geological era after ourselves), this is in tension with a broader recognition that humans are but one species of many, our survival is not guaranteed, and the inexorable march of geological time moves at a tempo not easily reconciled with or answerable to our human concerns. While these themes emerged to prominence in the 1970s and 1980s, as discussed in the previous chapter, by the 1990s and into the 2000s they had been reconfigured in dramatically different responses.

What most observers seem to agree on is that we have, in fact, crossed a threshold in which species loss and climate change have reached irreversible proportions, and human impact on the global environment is essentially indelible on any time scale meaningful to human beings. For some, this is a development essentially to be embraced—either as a challenge to human ingenuity to be solved by geoengineering, "de-extinction," space colonization, and the like, or as an opportunity to engage in a radical reevaluation of humanity's place in nature through development of a "multispecies ethics," the abolition of traditional categories of biological self-classification, and various forms of transhumanism. For others, though, it is a further indication that humanity is heading toward a catastrophic end. An intriguing aspect of this new apocalypticism, however, is that unlike late-nineteenth- and early-twentieth-century Modernists who saw the catastrophe as a possibly avoidable calamity, or Cold War pessimists who predicted an inevitable sudden holocaust, or postmodernists who viewed society as existing among the postapocalyptic ruins, the Anthropocene apocalypse is sometimes described as a "slow-motion catastrophe" that has been ongoing for decades, centuries, or even millennia. Our own chapter is unfolding in medias res, and while it may be hard to locate its beginning, it is similarly difficult to predict its endpoint: rather than envisioning a sudden fiery annihilation, we may have bequeathed a slow, protracted descent into greater misery and irrelevance to our future generations. Our current pessimists would argue that T. S. Eliot was right: *our* world, at least, may end "not with a bang but a whimper" (Eliot 1925).

Leaving such grand considerations aside for now, this final chapter will bring our narrative to a close by examining the formation of the science, rhetoric, and valuation of our most current version of the extinction imaginary. It has three concrete tasks to accomplish. The first is to document the emergence of the argument that we are currently

facing a crisis of declining biological diversity as the basis for a significant scientific movement. While concerns about the fate of endangered
species certainly long predated the 1980s, it was during that decade that
such anxieties crystallized into a broader mandate to preserve the diversity of *all* life. To put it another way, while past efforts tended to focus
on individual species under threat—and particularly on those to which
humans had some kind of emotional or economic attachment—the
biodiversity movement located its concern with diversity as a value in
itself. This frequently drew on arguments about the interconnectedness
of ecology, the dependence of human society on "ecosystem services"
(self-regulating properties of the organic and inorganic biosphere),
the importance of genetic diversity as a source for future evolutionary
potential, and the limitation of human knowledge about consequences
of drastic environmental change such as deforestation and global warming. In this formulation, the anathema was not just extinction but mass
extinction—defined as episodes during which a significant proportion
of the earth's species are lost during a sudden geological interval, with
perhaps a significant impact at the higher taxonomic levels as well. This
presented the prospect of long-term and irreversible ecological and
evolutionary trends.

A major source of information and rhetoric for biodiversity proponents came from the paleontological studies of mass extinction that rose
to prominence by the middle of the 1980s, as discussed in the last chapter. The study of extinction in life's past, it was argued, could be taken as
a model and a warning for understanding the present and predicting the
future, particularly in regard to the ecological dynamics that resulted
from sudden drops in life's overall diversity. While the resilience of the
biosphere was frequently noted—after all, life has recovered from the
five major mass extinctions of the past, and has even increased in overall taxonomic diversity—biodiversity champions were quick to point
out that such recoveries often took place on geological time scales (anywhere from five to twenty million years, depending on the severity of
the event) that dwarfed the span of our individual lifetimes and the entirety of human history itself. Furthermore, these recoveries, in both
the short and long term, have been highly unpredictable, and survival
and success has rarely been guaranteed to those species that formerly

dominated the globe. Human beings, after all, evolved from a lineage of small, insignificant mammals who benefited from the extirpation of the dinosaurs.

For this reason, it became rhetorically effective to compare the current depletion of diversity to past mass extinctions, and even to predict that anthropogenic species loss would eventually rival or exceed the greatest dyings of the past. The second section of this chapter will examine the development of this rhetoric, along with the scientific basis for analogizing between past and present mass extinctions. Ultimately, these arguments led to the formulation of the widely influential notion that we are currently witnessing a "Sixth Mass Extinction," first and most prominently advanced in Wilson's 1992 *The Diversity of Life* and subsequently taken up as a rallying cry for conservationists to this day. While the sixth-extinction concept has proven enormously effective in galvanizing public attention and concern, it has not been without its critics—including some of the very same paleontologists whose studies of mass extinction became so central to the conceit.

Finally, the chapter will conclude by examining the relationship between conceptions of biological diversity developed during the 1980s and early 1990s and a broader discourse of the value of diversity in the cultural sphere. By the early 2000s, a movement had emerged championing the protection of "biocultural diversity," in which the potential extinction of languages and other human cultural traditions was directly linked—through analogy—to the loss of biological species and genetic information. A central point to make here, though, is that the valuation of diversity of all kinds—and the threat posed to diversity by the specter of extinction—is a cultural and scientific co-construction. That is to say, while proponents of biocultural diversity drew explicit analogies between biodiversity and cultural diversity, this relationship was not *merely* analogical; as this book has argued, it is impossible to cleanly separate scientific values and beliefs from those circulating more widely in social, political, and cultural discourse. At least in Western society, the strong valuation that has, by our current moment, become attached to the inherent benefit of diverse forms of life, language, ideas, ethnicity, and other cultural forms is the expression of a deeper and more unified belief in the inherent value of diversity that exists prior to any specific disciplinary or cultural context.

From Endangered Species to Biological Diversity

On the final evening of the 1986 BioDiversity Forum, a group of the meeting's most prominent participants, including E. O. Wilson, Paul Ehrlich, Thomas Lovejoy (director of conservation at the World Wildlife Fund), and the botanist Peter Raven, convened a "national teleconference" to discuss the challenges facing the conservation of biological diversity. This event was broadcast live to more than one hundred colleges and universities, and was watched by an estimated audience of between five and ten thousand viewers.[5] During this teleconference, a passage was read from a statement issued during the forum by the so-called Club of Earth—a group of biologists including Wilson, Ehrlich, Raven, and others—arguing that the current extinction crisis was "a threat to civilization second only to the threat of thermonuclear war," a comment reported in several newspaper articles about the meeting.[6] The fact that this rather dramatic statement was not widely challenged in press accounts testifies to the rapid elevation of biodiversity as a broad political concern—as well as to the significant escalation in the stakes attached to conservation—during the 1980s. After all, despite high emotions attached to campaigns to protect endangered species such as the Siberian tiger, the California condor, and even the infamous snail darter, nobody ever claimed that the fate of the human species depended on their survival.

In truth, arguments around the preservation of species and ecosystems had taken an important turn beginning in the early 1970s, when in some quarters attention began to gradually shift from appeals for the protection of individual charismatic species, and toward stewardship of what would eventually be labeled "biological diversity." As historians have pointed out, a landmark moment in the establishment of the environmental movement in the United States—the passage of the Endangered Species Act (ESA) in 1973—built on growing momentum established during the previous decade and expressed by such highly visible public statements as Rachel Carson's *Silent Spring* and Ehrlich's *The Population Bomb*. The historian Mark Barrow, for example, notes that during this period, "Americans grew increasingly uneasy about myriad threats to their quality of life," including industrial pesticides and wilderness destruction, providing "fertile ground" for passage of the ESA (Barrow 2009, 348). What had previously been typically expressed as separate, if often politically aligned, interests—the protection of endangered species and the preservation of wilderness and natural environments—became joined, thanks in part to a growing recognition that human societies depend on natural resources that are bound together in complex ecological relationships. This is the "confluence of values" Farnham has described, which came together "in one concept that represented the protection of the living components of the natural world": the value of diversity. "By the 1970s," Farnham contends, "the desire to protect all of the natural variety present on Earth was most often expressed in conjunction with a reminder of all the benefits humans would lose should the diversity of nature be reduced" (Farnham 2016, 12).

This attitude is apparent, for example, in a report commissioned by the Committee on Science and Policy of the National Academy of Sciences in 1966, published several years later as a book titled *Biology and the Future of Man* (1970). Each chapter was composed by a panel of experts chaired by a prominent biologist, and the topic "The Diversity of Life" fell to the evolutionary biologist and systematist Ernst Mayr. This was significant because Mayr, whose considerable reputation derived in part from his activities in promoting the so-called Modern Evolutionary Synthesis of classical Darwinism with modern population genetics, gave a distinctly population-oriented spin to the problem of biological diversity.[7] While he noted that interest in natural diversity may be a kind of innate human inclination, he highlighted both the importance of species as "unique genetic system[s]," and the threat of extinction for reducing available genetic resources for future evolution. Arguing that "the important point is that the entire biota at any one time is interrelated and interdependent in an extremely complicated manner," Mayr warned that man's technological progress has released forces that lead to our ever accelerating destruction of natural habitats. Dozens, perhaps hundreds, of species are annihilated each year, species that required hundreds or thousands or millions of years to evolve. They cannot be replaced. Whenever man transforms the landscape for his own purposes, he destroys most of the native populations, usually causing their replacement by a few species that thrive in man-made environments.[8]

Two years later, an even more sweeping statement about the dangers of unchecked development was articulated by Barbara Ward and René Dubos in *Only One Earth* (1972), a summary of an unofficial report commissioned by the United Nations Conference on the Human Environment. Summarizing the views of a distinguished international panel including Konrad Lorenz, Peter Medawar, Margaret Mead, Jan Tinbergen, and the explorer Thor Heyerdahl (of *Kon-Tiki* fame), Ward and Dubos warned ominously that "the two worlds of man—the biosphere of his inheritance, the technosphere of his creation—are out of balance, indeed potentially in deep conflict." They argued that humanity stands at "the hinge of history," facing "a crisis more sudden, more global, more inescapable, more bewildering than any ever encountered by the human species" (Ward and Dubos 1972, 12). The nature of this crisis, they maintained, was the threat of widespread extinction triggered by the disturbance of finely balanced ecological systems. While they acknowledged that "interdependence of living things implies a certain stability," they nonetheless argued that "behind the interrelationships lies the risk of unpredictable and sometimes destructive consequences" that "can elicit so violent a response that the system may not be capable of returning, by itself, to a desirable and stable system" (Ward and Dubos 1972, 43). This risk was presented in the direst possible terms, as threatening not only the natural environments that humans depend on but the very future of humanity itself: "If man continues to let his behavior be dominated by separation, antagonism, and greed, he will ultimately destroy the delicate balances of his planetary environment. And if they were once destroyed, there would be no more life for him" (Ward and Dubos 1972, 45).

As these examples demonstrate, the notion that biological diver-

sity and ecological balance are vital resources for human civilization—
and are perhaps even essential to the continued survival of the human
species—was well established long before the biodiversity movement
of the 1980s and beyond took shape. Given its historical proximity to
other anxieties discussed in this book—the threat of nuclear war, the
pollution of the environment, overpopulation, and famine—concern
for the preservation of biological diversity should be considered a central part of Cold War extinction discourse, thoroughly intertwined with
these other fears. At the same time, however, discussions of biological
diversity presented an interesting new wrinkle, which manifested as a
tension between anthropocentric concern for diversity as a source of
essential resources and a broader ethical mandate to value the complexity of the natural world for its own sake. This tension has never been
resolved—indeed, it is one of the central features, and perhaps contradictions, of the current Anthropocene concept—but it would contribute substantially to the evolution of what would become the biodiversity movement.

In its basic formulation, the "resource" argument for maintaining
biological diversity has changed remarkably little over more than forty
years. Whether understood concretely as tangible material resources—
food products, medicines, economic goods—or more abstractly as "information"—for example, genetic information—the "utilitarian" value
of biological diversity has tended to take the spotlight. This notion was
enshrined in the justification for the 1973 Endangered Species Act,
which argued that, "from the most narrow possible point of view, it is
in the best interests of mankind to minimize the losses of genetic variations [whose value is] quite literally, incalculable [as] keys to puzzles
which we cannot solve, and [which] may provide answers to questions
which we have not yet learned to ask" (Congressional Research Service
1982, 144). More recently, the notion of biological diversity as an important contributor to "ecosystems resources" has been articulated to
describe the value of even the most humble species (such as bacteria,
algae, insects, and the like) to feedback mechanisms that regulate the
earth's water, soil, and atmosphere, on which humans depend.[9]

But from the very start, some conservation-minded biologists actively opposed anthropocentric-minded justifications for preserving

diversity, arguing that such a perspective served only to perpetuate attitudes that had brought the crisis on. One of the most prominent critics of anthropocentrism was the Rutgers University biologist David Ehrenfeld, a leader of the conservation biology movement during the 1980s, who made waves with his 1978 book *The Arrogance of Humanism*, which criticized what he called "the core of the religion of humanism: a supreme faith in human reason—its ability to confront and solve the many problems that humans face, its ability to rearrange both the world of Nature and the affairs of men and women so that human life will prosper" (Ehrenfeld 1978, 5). While this later book captured wider public attention, Ehrenfeld's first foray into the topic, his 1972 *Conserving Life on Earth*, helped set many of the terms for subsequent debate. In a striking analogy, Ehrenfeld described the attempt to convince the public to value the diversity of life on earth as being akin to "advertising color television on black and white screens," since "one can assert, persuasively, how beautiful and rich the colors are, but acceptance of the idea is still an act of faith on the part of the inexperienced audience" (Ehrenfeld 1972, xii). Dismissing traditional conservation efforts to preserve individual species as "elitist" and "pastoral," he argued instead for an ethic that was "holistic," acknowledging "both the complexity of ecological relationships and the high degree of connectedness binding together the biological world, the atmosphere, the surface of the earth, the fresh and salt waters, and the artifacts of human civilization" (Ehrenfeld 1972, 11).

Above all, Ehrenfeld argued, the "beast" or "central problem" facing humanity was "the loss of irreplaceable diversity," which he described as "outright theft, since once species have been obliterated they cannot be reconstituted" (Ehrenfeld 1972, 4). Indeed, *Conserving Life on Earth* was noteworthy for providing one of the first instances in print of the term "biological diversity," which Ehrenfeld described as "one of the main themes of this book" (Ehrenfeld 1972, 55). While he stressed that the concept was "naturally based in large measure upon *the number of species* in a given community," Ehrenfeld also acknowledged the importance of genetic diversity, protection of which was "a matter of retaining the maximum number of options for the future," or the maintenance of "irreplaceable biological 'information'" (Ehrenfeld 1972, 155).

On this basis, he argued, the "great tragedy of the Green Revolution"—
that is, of the agricultural initiatives during the late 1960s and 1970s that
saw particular, robust strains of wheat, rice, corn (maize), and other
crops established in Africa, Latin America, South Asia, and other areas
of famine and overpopulation, and which were widely credited with
staving off the dire scenarios predicted in Ehrlich's *Population Bomb*—
"is that it tends to destroy the very diversity that the world needs to survive and prosper" (Ehrenfeld 1972, 49). To Ehrenfeld, the unintended
consequences of intervening in nature were stark. In attempting to address one problem, well-intentioned planners had introduced another,
perhaps more severe: "the spread of a deadly agricultural uniformity."

For this reason, Ehrenfeld resisted the temptation to assign value
to nature as "resources," which he believed implied an "extractive" relationship towards nature (Ehrenfeld 1972, 9–10). This argument acquired even more force several years later with the publication of
The Arrogance of Humanism, where Ehrenfeld criticized "the humanistic world" for accepting conservation efforts "only piecemeal and
at a price, [demanding a] *logical, practical* reason for saving each and
every part of the natural world that we wish to preserve" (Ehrenfeld
1978, 177). Whereas his earlier appeal pointed to the tangible harm of
such practices in establishing agricultural monocultures, Ehrenfeld
now made the case for a new philosophy or ethic to guide conservation. This "conservative" value—by which he adamantly did not mean
the kind of conservatism normally associated with right-leaning political ideology—would explicitly oppose the "exploitative relationship
with Nature" often found in Western culture, since the preservation of
"non-resource" species was "often motivated by a deeply conservative
feeling of distrust of irrevocable change and by a socially atypical attitude of respect for the components and structure of the natural world"
(Ehrenfeld 1978, 178). Acknowledging that this view would strike many
as being "non-rational," Ehrenfeld argued that a new "value" had to be
constructed around the conservation of diversity. If for no other reason, it should have been apparent that species loss carried "a hidden
and unknowable risk of serious damage to humans and their civilizations," and that biological diversity must be preserved "because we do

not know the aspects of that diversity upon which our long-term survival depends" (Ehrenfeld 1978, 187–88).

These arguments about the valuation of natural biological diversity would have an important influence on the emergence of the field of "conservation biology" during the 1980s, a movement with which Ehrenfeld was closely associated, serving as the founding editor of the journal for the Society of Conservation Biology in 1987 (Soulé 1987, 4–5). They also had fairly immediate policy impact as well. As a by-product of the process that led to the establishment of the US Environmental Protection Agency by Richard Nixon in 1970, the Committee on Environmental Quality was established to report annually to the office of the president to assess and coordinate environmental and policy initiatives across federal agencies. In 1980, this annual report was presented to President Jimmy Carter, who officially submitted it to the US Congress with a short prefatory letter. While noting that significant progress had been made over the previous decade in controlling air and water pollution and encouraging alternative or more efficient use of energy resources, Carter's letter also sounded an alarm: despite encouraging evidence that the United States, at least, was moving towards sustainability, "there are also undeniable signs that in many other parts of the world the Earth's carrying capacity—the ability of biological systems to meet human needs—is being threatened by human activities" (Carter 1980, iii). If allowed to proceed unchecked, the letter continued, as many as "20 percent of all species of plants and animals on Earth, could disappear by the year 2000," a trend that could lead to "serious food scarcities" in many of the "poor nations of the world." The letter concluded, "We can no longer assume as we could in the past that the Earth will heal and renew itself indefinitely," since "humankind is now a potent force on the face of the planet. . . . The quality of human existence in the future will rest on careful stewardship and husbandry of the Earth's resources" (Carter 1980, iv).

While the report summarized initiatives and priorities across a wide variety of topics in economics, energy and natural resource management, land use, air and water quality, and environmental health, its first two chapters focused squarely on biological diversity. The first chapter, "The Global Environment," began with a warning that "a decline in the earth's carrying capacity" threatened resources "essential for human survival" such that the "capacity to support people is being irreversibly reduced." These included "essential" resources like water, fish, and timber, but also "hundreds of thousands of irreplaceable plant and animal species," especially in tropical forests (Council on Environmental Quality 1980, 1). The second chapter, "Ecology and Living Resources: Biological Diversity," highlighted the problem of extinction and the erosion of biological diversity in even starker terms. Noting that it is difficult to estimate current rates of extinction because many threatened species have likely never been identified and classified, the report warned of the possibility "that one to three extinctions are now occurring daily and that the rate will increase to one per hour by the late 1980s," resulting in a possible loss of as many as one million of the estimated "5–10 million species in existence worldwide . . . within our lifetimes." Such an event "would be unprecedented in the last 65 million years or, conceivably, since the beginning of life on this planet" (Council on Environmental Quality 1980, 31).

Framing the scope of this extinction problem with a rhetorical question, the report then asked why, "in a world filled with pressing problems . . . the loss of a million species should be considered an unparalleled tragedy." The "basic answer" it immediately supplied was "that by reducing biological diversity, humanity is squandering its greatest natural resource, on which we depend for food, oxygen, clean water, energy, building materials, clothes, medicines, psychological well-being, and countless other benefits" (Council on Environmental Quality 1980, 31). In the first instance, the value of biological diversity was presented squarely in terms of the language of resource: the "material value" of new sources of food, natural agricultural pest controls, untapped biological energy sources, chemicals and other raw materials, and of course pharmaceutical products. As the report put it, "In natural biological diversity, humankind has varied, infinitely renewable supplies of food, energy, industrial chemicals, and medicines" (Council on Environmental Quality 1980, 34). The report also stressed that these resources were not just material but also genetic, arguing that since "each species in a community is a unique genetic solution to a combination of environmental challenges," genetic diversity "maximizes the likelihood that at least some individuals of a species will withstand environmental change" (Council on Environmental Quality 1980, 33).

At the same time, however, the report also stressed reasons for preserving biological diversity that did not depend on material or economic considerations. Beyond an "ancient kinship" that humans feel with the natural world because of our shared evolutionary ancestry, the report cited philosophical, religious, and aesthetic arguments for preserving all living things, singling out Ehrenfeld's criticism of the limited persuasiveness of utilitarian arguments for particular mention (Council on Environmental Quality 1980, 40). Ultimately, the authors concluded that whatever the rationale, the best argument for protecting diversity is our own ignorance: since "our wisest contemporaries are those willing to admit the enormity of what is not yet known," any potential "discovery of the utilitarian values of the vast majority of species will lie in the future, if humankind allows them a future." In an echo of Ehrenfeld's critique of the unintended consequences of the Green Revolution, the report illustrated its case with examples of the harm caused by unstable monocultures introduced to address immediate problems, from the Irish potato famine of the nineteenth century to the recent introduction of hybrid rice strains in the Philippines. Given the importance of "genetic reservoirs to respond to fluctuating weather and rapidly evolving crop pathogens," the report's authors noted with wry irony the tendency for "modern agriculture . . . to kill the goose that lays the golden eggs" (Council on Environmental Quality 1980, 51).

It is worth pausing for just a moment to recognize the remarkable speed with which a notion that had been formally named perhaps barely a decade earlier had not only achieved a central place in a major US government report, but also had acquired a status of importance on par with other great global threats such as nuclear war, energy crisis, and pollution as a matter of pressing danger to humanity. It is true, as Farnham and others have noted, that previous conservation efforts, including the passage of the ESA, "opened the door" for interest in biodiversity, but that history alone does not explain the astonishing success that biological diversity had as a focus of scientific and political concern (Farnham 2016, 348). To adequately account for this transformation requires, as this book has argued, seeing concerns about biological diversity not just as a part of a history of conservation and environmental awareness, but against the broader backdrop of apocalyptic twentieth-century anxieties of all kinds, including fascination—both cultural and scientific—with catastrophic mass extinctions of the past and potential future.

"74 Species per Day": The Making of a Biodiversity Crisis

This last point suggests a question: If we grant that a "confluence of values" saw ecological theory, environmental activism, and political will coalesce successfully at a particular moment in history—1980 is a convenient date to locate this nexus—how did these environmental and biological concerns become central to the broader extinction imaginary developing at the time? In other words, what accounts for the ability of scientists to make—and journalists to uncritically report—a statement arguing that biological diversity loss is "a threat to civilization second only to the threat of thermonuclear war" only a few years later, and for the public and government organizations alike to take this seriously?

The very simple answer is numbers—but, as it turns out, the numbers are anything but simple. This point hinges on a matter both technical and rhetorical. From the technical standpoint, in order to demonstrate that a "mass extinction" is currently taking place, or is at least approaching, biologists needed some kind of quantitative metric to compare current biodiversity losses with the great episodes of mass extinction in the geological past. Helpfully, by the early 1980s paleontologists had provided some rough estimates (described in the previous chapter) of the percentage of families, genera, and species lost during the major extinction events at the end of the Permian, at the boundary between the Cretaceous and Tertiary (when the dinosaurs died out), and in other episodes of heightened extinction. As part of these studies, paleontologists had also attempted to calculate the normal "background" rate of extinction during periods of relative calm as a baseline against which to compare and identify mass extinctions. Furthermore, by the mid 1980s several paleontological studies—most prominently by Dave Raup and David Jablonski—examined the ecological and environmental consequences of past mass extinctions, determining, for example, the selective "rules" that apply following extinction events and the dynamics of ecological recovery in their aftermath. Though their authors were careful to acknowledge the great many uncertainties that factored into these estimates, such studies were often used to extrapolate to current biodiversity losses and their potential consequences. Indeed, paleontological evidence became a central pillar of the biodiversity movement as it evolved during the 1980s and 1990s.

From a rhetorical perspective, conservation advocates immediately realized the effectiveness of comparisons between the current biological diversity crisis and mass extinctions of the geological past. Ehrenfeld, for example, had argued in *Conserving Life on Earth* that "the current rate of extinction among most groups of mammals is approximately *a thousand times greater* than in the late Pleistocene, a geological epoch distinguished by a 'high' extinction rate" (Ehrenfeld 1972, 1972). Likewise, the *Environmental Quality* report of 1980 had concluded that potential species losses could reach 20 percent by the year 2000, a scale "unprecedented in the last 65 million years" (or, in other words, since the extinction of the dinosaurs; Council on Environmental Quality 1980, 31). And in his 1979 popular book *The Sinking Ark*, the biologist and environmental activist Norman Myers provided even more dramatic estimates, arguing that the next twenty-five years could see the extinction rate grow to forty thousand species per year, which "would amount to a biological débâcle greater than all mass extinctions of the geological past put together" (Myers 1979, 5). Such rhetoric proved enormously effective in attracting attention from the public and policy makers alike, and benefited greatly from the contemporary popular interest in mass extinction studies by Walter Alvarez and others. Over the next decade and more, these figures would in many ways come to define the biodiversity crisis itself. In his 1992 popular treatment *The Diversity of Life*, Wilson famously argued that a "cautious" estimate, "selected in a biased manner to draw a maximally optimistic conclusion, is the number of species doomed each year is 27,000. Each day it is 74, and each hour 3" (Wilson 1988, 280). Or, as he had put it a few years earlier in his opening keynote to the BioDiversity Forum, "The current reduction of diversity seems destined to approach that of the great natural catastrophes at the end of the Paleozoic and Mesozoic eras—in other words, the most extreme in the past 65 million years" (Wilson 1988, 11–12).

From this perspective, the claim that the biodiversity crisis rivals even the threat of nuclear war hardly seems excessive. Depending on how many species are estimated to currently exist, it might take only 100 years for current extinction rates to reach 50 percent or more of all life, and, as Jack Sepkoski calculated in a 1997 article, "only 355 years to eliminate 96 percent," a figure matching what is believed to have been the greatest of all mass extinctions at the end of the Permian (Sepkoski 1997, 536). It is indeed difficult to imagine the biosphere's recovery from such an event, much less the survival of human civilization.

The problem is that these figures were based on what were at best educated guesses about the number of currently existing species, the current rate of extinction, and extinction rates in the geological past. It should be emphasized here that I am in no way challenging the notion that biodiversity losses are significant, or that humanity faces a genuine crisis: whether it is one hundred, one thousand, or twenty-seven thousand species lost per year, it is still too many, and human beings bear the overwhelming responsibility for bringing on this state of affairs. What interests me as a historian, however—and what makes this issue particularly instructive for our broader survey of the history of extinction imaginaries—is the way that these figures and estimates found such a central place in political and scientific discussions of extinction, and how they reflect the longer history of anxieties about the future of humanity. As we will recall from the previous chapters' discussions of the threats posed by nuclear war, population explosion, nuclear winter, and other projected calamities, numbers and figures have been an essential component in creating broad acknowledgement, concern, and concrete action in relation to existential threats to humanity. During the 1950s and 1960s, for example, it was publication of stark facts and statistics about projected human casualties in a full-scale thermonuclear exchange that put a pin in optimistic claims for a "winnable" nuclear war, leading to politicians' embrace of a policy of nuclear deterrence. Likewise, the calculations of Sagan, Ehrlich, and others about the extensive aftereffects of a nuclear winter had a significant influence on efforts to deescalate tensions between the United States and the Soviet Union that ultimately led to the end of the Cold War.

The use of quantitative estimates of species loss, then, should be viewed as a further example of the power of statistics to convince the public to take heed and action in the face of events that might otherwise seem beyond the control of individuals or outside of the scope of human lifetimes. Statistics have long been a key weapon in scientists' arsenal for influencing popular opinion; this is simply a feature of modern science.[10] At the same time, statistical analyses tend to "black box" the phenomena they describe, often making them inaccessible to criticism or interpretation by members of the public or even other scientists who do not have access to the data or techniques relied on to produce them. A further aspect of black-boxing is that debate and discussion of contentious positions often takes place outside public view. Even if interested lay readers and policy makers theoretically have access to technical scientific journal articles, crucial debate can take place in correspondence between scientists, in informal discussions at meetings, and during the confidential peer review process. All of these factors contributed to the construction of the "biodiversity crisis."

In pointing this out, however, I do not mean to suggest that the biodiversity crisis is an example of unusual scientific practice or, more worryingly, a case of the improper imposition of "subjective" values onto science. In the first place—as this book has maintained from the very start—scientific arguments are never free from the values of the individuals and cultures in which they are framed. The notion of separate "scientific" and "cultural" spheres is, in my opinion, a misunderstanding of how science works. Science is part of culture, and while scientists employ tools, methodologies, and standards of evidence that are often different from other cultural productions (art or politics, for example), science is nonetheless a human production, and scientists are members of societies. One need only consider examples from the first two chapters of this book—concerning ideas about race or imperialism, for instance—to bear this out.

The false dichotomy between science and culture has unfortunately sometimes characterized commentary on the biodiversity movement. In his largely informative and instructive 1996 book *The Idea of Biodi-*

versity, the science studies scholar David Takacs makes the claim that "it is difficult to distinguish biodiversity, a socially constructed idea, from biodiversity, some concrete phenomenon," since scientists' "factual, political, emotional, aesthetic, ethical, and spiritual feelings are embodied in the concept of *biodiversity*" (Takacs 1996, xv, 2). So far so good, though I would argue that in this respect biodiversity is no different from most other scientific topics. But where Takacs's argument takes a wrong turn is in its further claim that "in so doing, scientists jeopardize the social trust that allows them to speak for nature in the first place":

> In the term *biodiversity*, subjective preferences are packaged with hard facts. . . . Biodiversity shines with the gloss of scientific respectability, while underneath it is kaleidoscopic and all-encompassing: we can find in it what we want, and can justify many courses of action in its name (Takacs 1996, 4, 99).

This view is problematic on two counts. Not only does it establish a false dichotomy between "subjective" and "objective" views of scientists, but it also mischaracterizes the debate itself as being far more nebulous than it actually was.

A scientist may well have personal reasons for pursuing a particular topic. A researcher in oncology may have lost a parent to cancer at an early age, or a Jewish physicist might have joined the Manhattan Project because she narrowly escaped persecution in Nazi Germany. These motivations may properly be considered subjective, but they are hardly determinative of the science produced. Closer to our case, virtually all researchers in natural history disciplines (e.g., botany, zoology, paleontology) report having been fascinated with, and even spiritually moved by, the beauty and complexity of nature from an early age.[11] This does not mean, however, that these scientists are necessarily compelled—consciously or unconsciously—to misrepresent the data or analysis of their subjects, or to attempt to mislead the public or their colleagues about their findings. They may—and often do—take up advocacy positions based on a combination of their scientific expertise and their personal values, but again this is hardly unusual. Prominent examples can be seen in virtually all scientific fields, from physics to biology, over the past two hundred years of professionalized science.

Takacs's argument about the biodiversity movement is based largely on the fact that, in his comprehensive survey of published literature and interviews with dozens of prominent biodiversity advocates (including Wilson, Ehrlich, and others discussed here), he discovered that definitions of what biodiversity is and how it should be valued varied quite widely. In identifying at least twenty-three such formulations of biodiversity, Takacs came to suspect that not only do scientists disagree about how biological diversity should be understood, but their definitions collectively encompassed such a range of features and values that the concept is rendered essentially meaningless: "Biodiversity's ecological value, therefore, looms inexpressively large, virtually unknown, but incalculably important" (Takacs 1996, 202). In other words, not only do biologists fail to agree on a basic definition for biodiversity, but they themselves are unable to articulate their own individual conceptions coherently and concretely, or to separate their personal values from their empirical conclusions.

But a central problem with this analysis is that Takacs has chosen to interrogate a nebulous concept to begin with. It is noteworthy that his book is titled *The Idea of Biodiversity* rather than *The Science of Biodiversity* or *The Politics of Biodiversity*. I suspect that one would encounter very similar disagreement, contradiction, and mixture of personal and collective values if one were to survey scientists for a study of "the idea of evolution" or "the idea of cosmology." The point is that if we set out by defining our categories in a way that does not distinguish between philosophical, personal, political, and empirical values and beliefs, we should not be surprised if we cannot disentangle them in our results. This is, in many ways, the approach the book you are reading has taken; it might well have been titled *The Idea of Extinction*, since it explicitly and intentionally seeks to understand the ways in which scientific discussions have been imbricated with cultural, political, and personal values. The difference, of course, is that I see this entanglement as essential to understanding how science works, rather than as a corruption of something that ought to be "pure."

In point of fact, biodiversity proponents were quite consistent about what biological diversity is, and about the threat its loss poses for the future. Central to the understanding of biological diversity as an empirical phenomenon and a resource—in literature from the early 1970s through the 1990s and beyond—is the perspective that ecology is an intricately interconnected system; that natural diversity, whether ecological or genetic, provides both resilience against sudden change and potential for future adaptation and evolution; and of course that mass extinctions, though part of life's natural order, have unpredictable and irrevocable consequences. This is not only a perfectly concrete and consistent conceptual stance; as this book has demonstrated, it is also the direct product of specific, contingent historical developments in the study of biology and paleontology. Furthermore, as this chapter will show, the measurement and assessment of the threat of biodiversity loss also followed a very consistent path: biodiversity loss is understood to be calculated by estimating the number of species extinctions in a given period (a day, a year, etc.) in relation to the number of species in existence, and the magnitude of the problem is calculated by comparing current rates of extinction to those in the geological past.

Quite importantly, most of the proponents of biodiversity conservation agreed both about the general estimate of current species extinction and about the potential for the crisis to escalate to levels approaching those of mass extinctions of the past. Disagreement, such as it was, came from disputes about the empirical basis for extinction projections owing to the poor state of existing taxonomic knowledge, especially for terrestrial invertebrates and plants in tropical environments. During the 1970s the generally accepted figure was that about 1.5 million species of plants and animals had been identified, but it was widely acknowledged that this number dramatically underestimated the true number of species alive—perhaps by one or more orders of magnitude. While conservative estimates placed the real figure at between three and five million species, many naturalists suspected that the number could be far higher, but at least ten million. For example, in his influential *The Sinking Ark*, Norman Myers based his projections of biological diversity loss on a figure of between five and ten million extant species, arguing that current extinction rates are at least one species per day. He predicted that this alone would be sufficient to alter "basic processes of evolution," but that the real danger lay in the escalating rate of species depletion, particularly through tropical deforestation (Myers 1979, ix). Myers acknowledged that his calculations—remember, he believed that the rate could ultimately climb to forty thousand per year—were based on a "guesstimate," but he warned that "any reduction in the diversity of resources, including the earth's spectrum of species, narrows society's scope to respond to new problems and opportunities," and that "the result will be a grossly impoverished version of life's diversity on earth, from which the process of evolution will be unlikely to recover for many millions of years." Ultimately, he predicted, "humanity might be destroying life that might just save its own" (Myers 1979, 7).

Although he never held a university professorship or similar position of institutional security, Myers became one of the most prominent and widely-cited figures in the unfolding biodiversity movement. Born in England, he spent his much of his early adulthood in Kenya, where he developed a love of natural diversity and a concern for threatened species and environments. After receiving a PhD in biology from Berkeley, he spent the rest of his career in a variety of short-term, often grant-funded positions, conducting ecological surveys and consulting on conservation projects for a variety of international agencies and foundations, including the World Wildlife Foundation (WWF); the International Union for Conservation of Nature (IUCN); the World Bank; and the national academies of the United States, the Soviet Union, and Sweden, as well as a plethora of initiatives and agencies sponsored by the US government and the United Nations. By the early 1980s he described himself as a "consultant in environment and development," presenting himself as an expert in topics including tropical forestry, human population expansion, energy resources, fisheries, land-use planning, and of course the economics and policy significance of biological diversity.[12]

Despite his hectic life (in an undated CV from the mid-1980s, he reported having worked on "more than 100 assignments in 40-plus countries" since 1982 alone), Myers was also a prolific author of popular books, essays, and scientific articles. Over his long publication career he authored or edited more than twenty books and nearly three hundred articles; during the 1980s, especially, he established himself as a widely cited expert on biological diversity issues. Along with Wilson and Ehr-

lich, Myers was one of the key figures in drawing both public and political attention to the biodiversity crisis, and *The Sinking Ark* is widely
considered to have had a formative impact on the conservation biology
and biodiversity initiatives of the 1980s. Yet he always viewed himself
as something of an outsider, and in numerous letters to Wilson during
the 1980s and 1990s he expressed anxiety about his lack of a stable position and source of funding for his research. Nonetheless, he and Wilson
developed a warm (though somewhat asymmetrical) friendship, with
Wilson frequently depending on Myers's wide-ranging conservation experience, and Myers on Wilson's contacts and clout in securing visiting
lectureships and awards.

This relationship began in earnest in 1976, when Myers wrote Wilson to seek "advice and assistance" during the planning stages for *The
Sinking Ark*. In particular, he wanted Wilson's views about the estimated number of species in existence, which he suspected might number more than ten million.[13] After the book's publication, Wilson ar-
ranged for Myers to visit Harvard (where Wilson spent his entire career,
from 1956 to 1996), and the two began a regular correspondence. Recalling their first meeting several years later, Wilson remarked that he
would "always recall the evening you lectured at Harvard, almost a lone
voice on the extinction problem, to an audience of perhaps 50 people,"
while "the rest of the university was off ogling that useless fool the Dalai
Lama," adding that "what you had to say was far more important than
anything His Holiness could say" (Wilson to Myers, July 19, 1983).[14] Wilson also frequently acknowledged the importance of Myers's work for
his own campaign for biodiversity. In 1980 he informed Myers that his
"1980s prophesy" (a short essay published in *Harvard Magazine*) "is
based on your important book *The Sinking Ark*." A decade later, he told
Myers he was "the most quoted author in *DOL* [Wilson's *The Diversity
of Life*, which was then in the prepublication stage]," and assured him,
"Your contributions will be showcased in this book" (Wilson to Myers,
December 30, 1991).[15] Wilson was indeed generous in crediting Myers's
contributions throughout his own publications on biodiversity, and
in arranging for opportunities for Myers to present his views—for instance, by giving him a prominent spot at the 1986 BioDiversity Forum.

Another central early influence on the biodiversity movement was

Paul Ehrlich, with whom Wilson had an equally close friendship, although one closer to a relationship of equals. Ehrlich, like Wilson, spent his entire career at an elite university (Stanford), where, like Wilson, he collected accolades, awards, and an endowed professorship while also writing (with his wife, the biologist Anne Ehrlich) a number of best-selling popular books. We have already discussed Ehrlich's role in sounding the alarm about global overpopulation and in developing the nuclear winter hypothesis, but at the same time he was also closely involved in raising awareness about the biological diversity crisis. Indeed, for Ehrlich these two issues were closely related. In 1981, he and Anne published a popular book very similar to Myers's *Sinking Ark*, titled *Extinction: The Causes and Consequences of the Disappearance of Species*. At Ehrlich's request, Wilson provided an effusive advance blurb, remarking, "*Extinction* is likely to be one of the most significant books of the 1980s, because it compellingly describes a phenomenon that may outrank even nuclear weaponry as the most profound long-term problem of mankind." In a private note to Charlotte Mayerson, a Random House marketing executive, Wilson commented, "You will be doing a major public service if you can turn the Ehrlichs' book into a best seller" (Wilson to Mayerson, February 26, 1981).[16]

Extinction opened with a striking metaphor that would be repeated often in subsequent public appeals about biodiversity. Imagine, the Ehrlichs asked, that a passenger jetliner were to lose a rivet from its wing. One or two missing pieces wouldn't affect the integrity of the plane, but at a certain threshold the entire structure would collapse, sending all of the passengers to their deaths. This was similar to the problem of species extinctions, they argued: "A dozen rivets, or a dozen species, might never be missed. On the other hand, a thirteenth rivet popped from a wing flap, or the extinction of a key species involved in the cycling of nitrogen, could lead to a serious accident" (Ehrlich 1986, xii–xiii). This "rivet-popping" metaphor came to signify a central plank of the biodiversity campaign: that small changes can have dramatic effects. (Similarly, Ehrenfeld had once likened this phenomenon to a grain of sand added to a gearbox.)[17]

Another feature of the Ehrlichs' argument was an explicit comparison between current species extinctions and those of the geological past. While they certainly employed this comparison for rhetorical purposes, they also used it to make substantive point: not only might the current mass extinction be as severe as past geological events in quantitative terms, but it might have even more serious evolutionary consequences. While mass extinctions of the past had indeed removed substantial portions of the earth's existing diversity, they nonetheless also left sufficient genetic and ecological resources for diversification to recover and thrive. The Ehrlichs argued, however:

> Extinctions that are occurring today and that can be expected in the future are likely to have much more serious consequences than those of the distant past. First of all, unless action is taken, contemporary extinctions seem certain to delete a far greater proportion of the world's store of biological diversity than did earlier extinctions [largely because they are taxonomically more widespread, whereas previous extinctions tended to differentially impact a smaller number of higher taxa]. Furthermore, the same human activities that are causing extinctions today are also beginning to shut down the process by which diversity could be regenerated. Entire new groups of organisms are unlikely to evolve as replacements for those lost if Earth's flora and fauna are decimated now (Ehrlich 1986, 10).

This last argument addressed one of the features of mass extinctions in the geological past discussed in the last chapter: the sense in which mass extinctions are, as Raup, Jablonski, Sepkoski, and other paleontologists stressed, "constructive" as well as "destructive" events from an evolutionary perspective. That constructive aspect can only act on environments that continue to be physically hospitable to life, and on ecosystems that retain sufficient genetic diversity for natural selection to produce new adaptations for changed conditions. The significant current extinction of plants, for example—which had come through relatively unscathed in previous extinction events—was particularly worrisome, since this suggested a potential breakdown of chemical and energy cycles on which all life depended.

Regarding the values attached to biological diversity itself, the Ehrlichs rehearsed arguments that were by now becoming familiar. Genetic variability was described as a source of resistance to extinction in the face of environmental change, while ecological diversity was described as "an enormous organic 'library' from which humanity has already drawn a vast array of useful substances" (Ehrlich 1986, 90). Here the Ehrlichs made it clear that, unlike Ehrenfeld, they regarded utilitarian arguments as being most compelling for protecting biological diversity. While they acknowledged that valid reasons for valuing diversity included "simple compassion" or "beauty, symbolic value, or intrinsic interests," they highlighted as "the most important of all the arguments" the fact that "other species are living components of vital ecological systems (ecosystems) which provide humanity with indispensable free services—services whose substantial disruption would lead inevitably to a collapse of civilization" (Ehrlich 1986, 6). Indeed, throughout the early biodiversity campaign, Paul Ehrlich would be one of the most outspoken champions of the "ecosystem services" argument, a notion he and Anne had first promoted as early as 1970 in their textbook *Population, Resources, Environment* (Ehrlich and Ehrlich 1970, 157).

Ehrlich's unabashedly anthropocentric view was not, however, the only important perspective in the emerging biodiversity movement. In 1985, the broad-spectrum biological magazine *Bioscience* featured a special issue titled "The Biological Diversity Crisis," with articles by a number of biologists and ecologists including Wilson and Michael Soulé. Soulé, who would be instrumental the next year in establishing the Society of Conservation Biology and its flagship journal *Conservation Biology*, took the opportunity to use his article "What is Conservation Biology?" to promote the new subdiscipline as "a new stage in the application of science to conservation problems" that "addresses the biology of species, communities, and ecosystems that are perturbed, either directly, by human activities or other agents." Stressing that "its goal is to provide principles and tools for preserving biological diversity," he argued that "it is often a crisis discipline," meaning that "one must often act before knowing all the facts; crisis disciplines are thus a mixture of science and art, and their pursuit requires intuition as well as information" (Soulé 1986, 727).

One of Soulé's central arguments was that, unlike "natural resource fields" that deal with the economics and other practical aspects of environmental regulation and policy, conservation biology is not primarily concerned with "utilitarian, economic objectives" or with preserving only "a small number of particularly valuable target species" (Soulé 1986, 728). Rather, he urged, the field should take a "holistic" view to preserving all forms of life. This he justified by presenting two sets of "postulates," which he divided between "functional" and "normative" considerations. In the first case, Soulé emphasized "evolutionary" features of ecosystems, which stressed the ecological interdependence of species on one another as a bulwark against extinction: issues of "scale" or thresholds above and below which ecological processes "become discontinuous, chaotic, or suspended"; and "population phenomena," such as the influence of natural selection, genetic drift, and population size on ecological stability (Soulé 1986, 729–30).

The second set of "normative" postulates were essentially expressions of the values that Soulé believed inevitably followed from the functional ones, the first being that "diversity of organisms is good." Perhaps surprisingly, he explained that "such a statement cannot be tested or proven," but that it may reflect some deeper human instinct to "enjoy variety" (Soulé 1986, 730). Soulé's postulates were not intended to be arguments based on empirical evidence or logical deduction. Rather, like the postulates of Euclidean geometry, they were claims understood to be self-evidently true, forming the starting point for further argumentation. A "corollary" of the inherent value of diversity was, according to Soulé, that "the untimely extinction of populations is bad," although he was quick to note that "conservation biology does not abhor extinction per se," since in its "natural" form "it is part of the process of replacing less well-adapted gene pools with better-adapted ones." The essential point, though, was that "natural" extinctions (Soulé seems not to have been troubled by the vagueness of this term), understood to be rare events, did not reduce biological diversity, since they were "offset by speciation." It was only when they took place in "catastrophic" fashion, as in the current biological diversity crisis, that they upset the natural ecological balance.

A critic might point out that in defining the values of biological diversity as self-evident "postulates" while justifying them with reference to so-called natural processes like natural selection and evolution, Soulé was trying to have his cake and eat it too. That is to say, he was attempt-

ing to ground the value of biological diversity in basic principles of evolution and ecology (the "functional" postulates), while simultaneously
implying that they were intrinsic and needed no prior justification. In
any event, he defined several additional postulates, including the assertion that "ecological complexity is good" (because it maintains habitat
diversity and stability), that "evolution is good" (because it is the "machine" of diversification), and ultimately that "biotic diversity has intrinsic value, irrespective of its instrumental or utilitarian value" (Soulé
1986, 731). This final normative postulate was "the most fundamental,"
Soulé argued, since "in emphasizing the inherent value of nonhuman
life, it distinguishes the dualistic, exploitative world view from a more
unitary perspective: species have value in themselves, a value neither
conferred not revocable, but springing from a species' long evolutionary heritage and potential or even from the mere fact of its existence."

Soulé evidently had sent a draft of this essay to Wilson for comment, since more than a year prior to its publication Wilson had written
to say, "I like your essay 'What is conservation biology,' as I have liked
most of your writings, as well as appreciated your pioneering role in
creating conservation biology." Wilson also commented approvingly on
the term "crisis discipline," which he regarded as a "valuable concept"
for promotion of the field (Wilson to Soulé, August 31, 1984).[18] Wilson noted that he would forebear commenting on the essay "at length,"
since his forthcoming book *Biophilia*, to be published later that year,
was "a lengthy commentary on most of the topics you raise. . . . We are
indeed thinking about the same things." He added that "the *crucial* step
is getting these issues on the national agenda" would require both a "literary" approach to reach the general public and a "political" one. Here
he noted his activities with the international development board at the
NRC: "With more of us pushing in the same direction, movement may
result."

In the end, Wilson's essay "The Biological Diversity Crisis" was published in the same issue of *Bioscience* as Soulé's, supplementing Soulé's
philosophical arguments about the values of biological diversity with a
detailed empirical accounting of the scope of the problem, along with
specific policy recommendations.[19] Somewhat mysteriously (I have
found no explanation either in print or in private correspondence), this

essay was also published, in nearly identical form, in another general science journal, *Issues in Science and Technology*, at virtually the same time as it appeared in *Bioscience*.[20] It appears that Wilson simply wanted his arguments to reach the widest possible readership. In a letter to the managing editor of *Issues*, he explained, "I am anxious to bring the subject to the attention of a broader audience. . . . Biological diversity is one of the rapidly emerging but still poorly articulated issues" (Wilson to Cook, June 14, 1985).[21] At around the same time, Wilson thanked his colleague Peter Raven—the director of the Missouri Botanical Garden and an important biodiversity advocate—for commenting on his manuscript, similarly explaining its purpose as "get[ting] the problem, particularly that concerning systematics, before as large and influential audience as possible in a form that will be read and remembered" (Wilson to Raven, June 10, 1985).[22] Wilson went on to observe, "A sea change may be in the making. . . . You, Norman Myers, and a very few others deserve a great deal of credit [but] a lot remains to be done." Putting biological diversity on the radar of influential politicians, he argued, was of the utmost importance: "The important people in Congress and elsewhere know all about bioengineering, nuclear winters, and the population bomb. . . . Now it's just a matter of getting tropical deforestation and the diversity crisis on the top-level agenda. And hopefully with as positive, upbeat tone as can be mustered." While it was not explicitly mentioned in his letter, the BioDiversity Forum to be held the following year, with which Raven was closely involved, was designed precisely to achieve this goal.

Wilson's essay in *Bioscience* presented a set of arguments that would be repeated in most of his subsequent appeals for biological diversity preservation, and in particular it emphasized the quantitative dimensions of the crisis. The essay began with the observation, "Certain measurements are crucial to our ordinary understanding of the universe"— such as the diameter of the earth, the number of stars in the Milky Way, and the mass of an electron. To these, Wilson added the number of species currently alive (Wilson 1985, 700). The problem was that, unlike those other figures, biological diversity had not been adequately measured, "not even to the nearest order of magnitude." Like Myers and others before him, Wilson noted that current tabulations of exist-

ing species—around 1.7 million, according to the most recent count—
"grossly underestimate[d] the diversity of life on earth," despite the best
efforts of systematists to improve the state of knowledge. The only way
to begin to assess the scope of the diversity crisis was to rely on estimates and extrapolations—to make a "guesstimate," as Myers had put
it in *The Sinking Ark*.

Unlike Myers, the Ehrlichs, and other previous commenters, however, Wilson was armed with a new source of information. In 1982, in a
short paper in the relatively obscure journal *The Coleopterists Bulletin*,
the entomologist Terry Erwin had extrapolated the "true" number of
arthropod species globally from a sample of a single hectare of Panama-
nian tropical forest (Erwin 1982, 74–75). Though it clocked in at a mere
five brief paragraphs over two pages, Erwin's article would prove to be
massively influential for the biodiversity movement, and ultimately can
be considered the source of many dramatic claims for current extinction rates presented over the next decade—including Wilson's figure of
seventy-four species per day in *The Diversity of Life*. Erwin's argument
was simple and elegant. It began with the familiar observation that,
while for more than a century naturalists had speculated that the vast
majority of living arthropod species remained unclassified, estimates
nonetheless put the true figure at between 1.5 and 10 million species.
Since this uncertainty largely owed to the fact that arthropod species
tended to be found in small, locally endemic populations in inaccessible
places, such as tropical forests, Erwin argued that if one could estimate
the actual number of species—including unclassified ones—in a single
local area, a more reliable figure could be extrapolated for the global
population.

Erwin did not actually count the total number of species of arthropods in the single hectare of tropical forest he chose. He limited his
census to only those species found in the upper canopy of the forest,
where animals could be collected from the "large and wide-spaced
leaves" of the tree *Lueha seemannii* (a kind of evergreen with large,
flat leaves, found in Central and South America) (Erwin 1982, 74).
Over three seasons of sampling, Erwin's team identified more than 955
species of beetles alone, to which he added another 206 weevil species
(identified in similar surveys in Brazil). Then came the extrapolation:

since it was known that up to 245 species of trees could exist in a single hectare of "rich" tropical forest, Erwin deduced that an average of 70 separate tree genera was reasonable. Next, he estimated the number of "host-specific" arthropods that occupied each genus (i.e., the number of arthropod species adapted to one and only one genus of tree), which he "conservatively" put at 20 percent of the total census. Using these estimates, Erwin calculated that his single case-example of *Luehea* carried "an estimated load of 163 species of host-specific beetles," the rest of which were "transient" (i.e., "resting or flying through *Luehea* trees"). If there were 70 genera of "generic-group tree species" in a hectare of forest, this would mean that there were "11,410 host-specific species of beetles per hectare, plus the remaining 1,038 species of transient beetles, for a total of 12,448 species of beetles per hectare of tropical forest *canopy*" (Erwin 1982, 75). Erwin did not stop there: noting that beetles compose 40 percent of all arthropod species, he reasoned that this would imply that the total number of arthropods in a hectare of tropical forest canopy was a staggering 31,120; and since he believed that the canopy fauna was "twice as rich as the forest floor and composed of a different set of species for the most part," this meant that adding another multiple of one-third to the number would produce "a grand total of *41,389 species per hectare* of scrubby seasonal forest in Panama!" (Erwin 1982, 75). Applying this formula to the "estimated 50,000 species of tropical trees," and assuming that "tropical forest insect species, for the most part, are not highly vagile [i.e., don't move around much] and have small distributions," Erwin concluded that "there are perhaps as many as *30,000,000* species of tropical arthropods, *not* 1.5 million."

Erwin's estimate is indeed staggering, and it suggests that past estimates of the diversity of life may have been low by as much as two orders of magnitude. Wilson cited Erwin's study as the source of his own reasoning in "The Biological Diversity Crisis," using it as the basis for his estimate of the influence that tropical deforestation is having on extinction, since it should now be possible to approximately calculate what the reduction in tropical habitat—a figure easily obtained from geographical surveys—would have on the loss of biodiversity.[23] Wilson freely acknowledged that such estimates should be bolstered by more accurate assessments of actual species numbers, and the remainder of his article argued for greatly increased funding for classification projects. Still, if Erwin's numbers were even remotely accurate, the true scope of the extinction crisis was magnified enormously. Instead of losses of perhaps several hundred tropical species per year, we are in fact dealing with a scenario in which annual extinctions could reach the tens of thousands.

That is, of course, a significant "if," and a number of scientists jumped into the debate to raise concerns. For example, in a 1988 article titled "How Many Species Are There on Earth?" the Australian ecologist Robert May—a pioneer in the field of theoretical ecology and a highly decorated professor at Oxford University—argued that Erwin's estimate "has been widely cited often without full appreciation of the chain of argument underlying it" (May 1988, 1448). As May pointed out, the assumption that 20 percent of beetle species are found on only one species or genus of tree is entirely arbitrary, as is the two-to-one ratio of canopy arthropod species to ground species. Furthermore, we simply don't know what percentage of canopy fauna is beetles; a survey of a single hectare is insufficient to draw significant conclusions. It might further be argued that there are other shaky assumptions in Erwin's analysis: he obtains the figure of 30 million species by dividing the total number of canopy species (31,120) by the average number of species and genera of trees per hectare, another fairly arbitrary assumption (70), thus producing an estimate of roughly 444 species that can be attached to a given species of tree. This is then multiplied by the estimated 50,000 species of tropical trees (again, an estimate based on the assumption that many species of tropical trees have not been discovered), which produces 22,200,000 canopy species. To this Erwin adds one-third of the total number, because of the two-to-one ratio of canopy to ground species—another 7,326,000—for a grand total of 29,526,000.[24] But why assume that arthropod species endemic to a particular tree species in one area are not found in another tree species elsewhere, especially if the floral composition is different in the two regions? Why assume that the average number of arthropod species per tree average is stable across all regions of the tropics? Why assume that the number of ground and "transient" arboreal arthropods—which are not tied to specific tree species—can simply be multiplied by 50,000 along with the endemic species (this is an implicit but unstated assumption in deriving one-third of the 22 million endemic arboreal arthropods as additional distinct species)? And why indeed assume, as May wondered, that beetles account for 40 percent of all undiscovered canopy-dwelling arthropod species?

The truth is that a significant pillar of the empirical estimate of biodiversity loss—and one that has remained in circulation for decades—depends on what are, even by Erwin's own admission, highly speculative numbers. As Erwin put it in 1982, "I would hope someone will challenge these figures with more data." This is precisely why Wilson argued for the necessity of massive diversity surveys. But at the same time, and in the unfortunate absence of significant resources to fund basic systematics research, Erwin's figures continued to bolster claims about extinction rates, whether or not Erwin's study was explicitly acknowledged. This is not, in any way, to minimize the seriousness of the biodiversity crisis: even May acknowledged that "maybe half of all extant species will become extinct in the next 50 or 100 years if current rates of tropical deforestation continue" (May 1988, 1448). But biologists have no idea whether that actual number is in the thousands or the millions. In a critical evaluation of another manuscript by Wilson—a paper presented in 1986 at a joint meeting of the Royal Academy of London and the American Philosophical Society, titled "Biological Diversity as a Scientific and Ethical Issue"—Raven made several of these points to Wilson. He cautioned Wilson, for example, "I don't think you have any real reason for saying that the absolute number of insects certainly exceeds five million, and I don't think anyone has really investigated the basis for Terry Erwin's estimates carefully" (Wilson to Raven, June 6, 1986).[25] He also expressed concern that his own work with Wilson on a 1980 report, arguing that there were "twice as many kinds of organisms in the tropics, minimum," had "loosely and without any particular foundation slid up to 'everyone agrees that there are at least 5 to 10 million species of organisms in the world,'" and that this was "a sort of non-conservative, very loose and non-scientific kind of estimate which is being accepted primarily by repetition."

In his reply, Wilson conceded some of Raven's points but defended

his basic reasoning, in part on the basis that "my more generous estimates illustrate zoologist [Wilson]-vs.-botanist [Ravin] once again. . . .
I think I'm right this time" (Wilson to Raven, June 10, 1986).[26] He based
this largely on personal anecdotal experience, noting:

> Time and again, sometimes to my dismay, I have watched an ant "species"
> dissolve into 2, 3, or more undeniable sibling species. . . . Most ento-
> mologists [working in the tropics,] even casually, have stories of 20 new
> species of such-and-such beetle or thrips genus discovered on one tree
> species, 8 new species of mites found in one berlesate, and so forth. The
> overall impression is one of a huge fauna of which only a small fraction
> is yet known. So at least keep in mind that insects *are* different, and pos-
> sibly some other invertebrates as well. Thirty million may well be far too
> high, but 5 million isn't.

But Wilson also stressed that differences in empirical calculations
shouldn't fundamentally affect the plan for taking concrete action, and
he defended the use of estimates while more concrete data were still
unavailable. Both he and Raven supported a comprehensive global survey, and in the meantime it was vital to "get [policy makers'] attention
with striking and defensible facts," since "when enough people of influence care about the problem, they can be presented with detailed
procedures and solutions." After all, Wilson concluded after he presented the paper at the Royal Society/APS meeting, "no less a person
than [the Nobel Prize–winning economist] Milton Friedman rose to say
that tropical deforestation should now be regarded as a global problem
comparable to the threat of nuclear war. Now *that* is a piece of tangible
progress, enough to keep me going."

The Sixth Extinction

Just a few months after the exchange between Wilson and Raven described above, the two sat together onstage during the national teleconference following the BioDiversity Forum, when the biological diversity crisis was proclaimed to be "a threat to civilization, second only

to the threat of thermonuclear war." It is impossible to know whether Friedman's comment inspired Wilson to make this claim in such a public forum, but evidently, despite their empirical disagreements, Wilson and Raven could agree on the general magnitude of the crisis. Indeed, following the forum, the biodiversity movement gained rapid momentum. Most of the papers delivered at the Washington meeting were published two years later in a hefty volume sponsored by the National Academy Press, and in 1991 the journal *Science*—probably the most widely read and respected general scientific journal in the world—made space for a special issue on biodiversity featuring articles by Ehrlich and Wilson, Soulé, Erwin, and Jablonski (who provided a paleontological perspective).[27] On the political front, the movement had a stunning success in 1992 with the adoption of the UN Convention on Biological Diversity, which asserted "the intrinsic value of the ecological, genetic, social, economic, scientific, educational, cultural, recreational and aesthetic values of biological diversity and its components." Underlining the success of the rhetorical and scientific arguments made by Wilson and his colleagues during the previous decade, the convention further acknowledged "the importance of biological diversity for evolution and for maintaining life systems of the biosphere" and affirmed that "the conservation of biological diversity is a common concern for humankind" (United Nations 1992).

As successful as Wilson and others had been in establishing biodiversity as a central scientific concern, they still needed to reach the public with their message. As we have seen in examples such as "mutually assured destruction," the "population bomb," and "nuclear winter," it helped to have a catchy slogan to attract widespread attention, and to galvanize popular interest and political action. For the biodiversity movement, that slogan would be "the sixth mass extinction," or just "the sixth extinction"—a term that by the early 2000s would become thoroughly entrenched in both the popular and the scientific discourse around biodiversity. The very first published instance of the exact phrase seems to have been in Wilson's *The Diversity of Life*, where at the conclusion of a chapter outlining the scope of the biodiversity crisis he opined, "Humanity has initiated the sixth great extinction, rushing to eternity a large fraction of our fellow species in a single generation"

(Wilson 1992, 32). Of course, as we have seen, a number of observers, including Wilson, had made explicit reference to past mass extinctions when discussing current biodiversity depletion over the previous decade. And in a 1988 essay for a special publication sponsored by the National Geographic Society titled *Earth '88: Changing Geographic Perspectives*, Wilson noted, "Virtually all students of the extinction process agree that biological diversity is in the midst of its sixth great crisis, this time precipitated entirely by Man" (Wilson 1988, 76).

But I argue that the notion of a "sixth extinction," more than acting as a catchy slogan, represents an important development in the conceptualization of extinction as an ecological process, as well as a major shift in the wider cultural discourse linking the science of mass extinction to broader concerns and anxieties around the fate of humanity. As a mode of "extinction discourse" or "catastrophic thinking," the sixth extinction is the final stage in the main narrative of this book; and analyzing its emergence sheds light on the way many observers have come to see our own current moment as a distinct stage in geological history. If late Cold War culture was characterized by a set of fears tied to the threat of sudden, catastrophic annihilation through nuclear Armageddon—a kind of secular apocalypticism that resonated deeply with the linear, progressive narrative of Judeo-Christian sacred history—then the sixth extinction and the Anthropocene concept signify something new about Western culture's imaginary of deep time. Mass extinctions, now understood to be a regular feature of the history of life, suggest an alternative to a narrative in which the emergence of human civilization is the alpha, and its potential, perhaps even inevitable self-destruction is the omega. As discussed in the last chapter, many paleontologists observed that the so-called "new catastrophism" of the 1980s was also a kind of "new uniformitarianism": viewed from the perspective of geological time, the history of life resolves itself into a series of crises, spaced fairly regularly, that imply an element of cyclicity underlying the more directional trends seen at lower levels of temporal resolution.[28] The perspective implicit in the sixth extinction and Anthropocene concepts likewise situates the directional, contingent human story within the broader cycles of geological time. Humanity may have cast itself in the unusual role of both "asteroid" and "dinosaur" in the impending environmental crisis,

but it now does so with a resigned awareness that even the most catastrophic result in the short term will be subsumed into the deeper cyclical patterns of natural history.[29]

Of course, this does not necessarily mean acceptance of a sixth mass
extinction as a foregone conclusion, nor do I mean to imply that biologists and paleontologists regard humanity's destructive impact on
the earth as a "natural" or excusable phenomenon. But if an implicit
assumption in framing the biodiversity crisis as a sixth extinction is that
the current depletion of biological diversity can best be understood
in—and its consequences extrapolated from—the context of the "Big
Five" mass extinctions of the past, there is a sense in which anxieties
about humanmade catastrophe have become naturalized by association
with broader natural cycles of deep time. Indeed, while this has been
a persistent feature of biodiversity rhetoric since the mid-1980s, it has
also been a relatively underexamined one. The route from biological diversity crisis to sixth mass extinction appears, as discussed above, as a
fairly unproblematic syllogism in the writings of Myers, Ehrlich, Wilson, and other biologists through the 1980s and into the 1990s. Estimates of current biodiversity loss suggest a far greater extinction rate
than the normal "background" rate of extinction in geological time; if
allowed to proceed unchecked for decades, the quantitative species loss
could rival the total estimated loss during the Cretaceous-Tertiary mass
extinction. The biodiversity crisis is therefore a "mass extinction," and
can best be understood by comparing its dynamics to those mass extinctions that have been studied in the deep history of life.

There is nothing inherently objectionable about this logic, aside
from the fact that its empirical evidence relies on some debatable figures, such as Erwin's estimates of tropical invertebrate endemism. But it
does require some closer examination, particularly in its conclusion. It
is a virtual certainty that humans are causing extinctions at a rate much
higher than would normally obtain during periods of environmental
calm; this is true whether or not we accept Erwin's estimates or Wilson's extrapolations. But it is not necessarily the same thing to observe
that we are experiencing short-term species loss in particular ecological niches and among specific groups of organisms, and to claim that a
"mass extinction" is taking place—particularly in the very specific sense

in which paleontologists developed the understanding of mass extinctions (and their ecological and evolutionary consequences) during the 1980s. Again, my point here is not to deny that the biodiversity crisis is real and demands action. Rather, it is to unpack some of the historical circumstances in which the sixth extinction concept was framed, to examine the relationship between the claims made by biodiversity proponents and the views of the scientists on whom these claims often depended, and to suggest some potentially unexamined and unintended consequences of the rhetorical and conceptual move from "crisis" to "mass extinction."

In the first place, it should be stressed that the sixth extinction concept underlines how closely the biodiversity movement depended on new understandings of extinction and diversification developed by paleontologists over the previous few decades. This is a fact that has not been fully appreciated by the literature on the biodiversity crisis, but it is central to understanding how and why biological diversity became a topic of such central concern when it did. While ecologists and conservationists had expressed long-standing concern about the fate of specific endangered species, it was the paleontological perspective on coordinated mass extinctions that emerged during the 1980s which shifted the focus to the potential loss of entire ecosystems and the protection of biological diversity per se. Prior to the late 1970s, many— if not most—biologists doubted whether mass extinctions could take place at all, such was the assumed resilience of the "balance of nature." Paleontological studies demonstrated, via detailed empirical investigations of particular stratigraphic breaks as well as broad statistical analysis, not only that mass extinctions have been a regular feature of the history of life, but also that they had long-term ecological and evolutionary consequences for the future diversification of life. In a sense, the very idea of "mass extinction" was constructed by paleontologists such as Raup, Sepkoski, and Jablonski during the 1980s.[30]

Secondly, paleontological mass extinction studies—and in particular the Alvarez team's hypothesis of the asteroid impact that wiped out the dinosaurs—stoked public interest in mass extinction and created a receptive environment for appeals by biologists and ecologists to present arguments about the current biodiversity crisis. Despite decades of environmental activism, by the late 1970s conservation organizations like the WWF and the Sierra Club had run into significant public apathy and political pushback around endangered species protection. The election of Ronald Reagan to the US presidency in 1980, and his appointment of divisive figures such as Interior Secretary James Watt, caused genuine concern that gains made over the previous decade in environmental policy and legislation could be rolled back. Indeed, in 1981 and 1982 the US Senate conducted a series of hearings to determine whether the Endangered Species Act, set to expire in 1983, should be reauthorized, thus causing a significant public stir. As the *New York Times* reported at the time, many government scientists were concerned that the Reagan Administration had failed to support the provisions of the ESA, and had in fact gone so far as to remove a number of threatened species from protected status. Citing the harmful influence of Watt as interior secretary, one anonymously quoted scientist complained, "Not one new species has been listed since Reagan came in. Nothing" (*New York Times* 1981). The article went on to note environmentalists' opposition to "a new priority system" that "concentrates resources for saving the most endangered species among the higher orders of life, such as mammals and birds," noting Environmental Defense Fund activist Michael J. Bean's concern "that the preservation of the earth's genetic diversity required that the protection of the act be extended to all life." In fact, Bean was in correspondence with E. O. Wilson at the time, encouraging Wilson to testify before the Senate subcommittee (he did) and requesting Wilson's endorsement of a statement on biological diversity cosigned by G. Evelyn Hutchinson, Peter Raven, and Thomas Lovejoy.[31]

The point here is that by the early 1980s, the political landscape around environmental protection in the United States had changed quite dramatically from the previous environment-friendly administrations of Carter and Nixon during the 1970s. (The bipartisan nature of environmental concern during the 1970s also highlights the dramatic political shift that took place during the 1980s in the United States.) Ultimately, while the ESA was reauthorized, Wilson and others realized that in order to guarantee continued public and political support for endangered species protection, the focus would have to shift away from the individual species (the infamous "snail darter" episode underlined the danger of relying on public enthusiasm for protecting endangered species that were not "charismatic") and toward economic and environmental arguments for protecting biological diversity itself. But convincing a skeptical audience that the depletion of diversity could have serious consequences required presenting that threat in concrete terms. Fortunately, the Alvarez hypothesis and related paleontological studies of mass extinction provided a ready-to-hand and highly visible scenario for conveying these consequences to a wide public. Rather than having to explain in detail the complex ecological basis for ecosystem interdependencies, Wilson and others could, if necessary, simply draw an analogy with the spectacular mass extinction events of the past. And, given the close association at the time between the presentation of the dinosaur extinction and nuclear winter scenarios, mass extinction was a potent threat indeed.

The role of paleontologists themselves in the biodiversity movement is a somewhat complicated story. A number of paleontologists—including most prominently David Jablonski, Jack Sepkoski, and Niles Eldredge—were early proponents of linking historical studies of mass extinction to the present crisis, and since the 2000s a number of other paleontologists have endorsed the notion that current biodiversity loss is contributing to a sixth mass extinction.[32] At the same time, some extinction experts—most notably Dave Raup—have questioned the analogy between past and present extinction rates; and even those generally supportive of the claim that we are experiencing a mass extinction event have expressed reservations about some of the ways in which paleontological data have been used in these arguments. At the heart of the issue is a question about whether estimates of extinction rates in the geological past are a valid basis for extrapolation to the present. Nearly all calculations of the magnitude of the current crisis by Wilson, Ehrlich, Myers, and other biologists depend on Raup's estimate that the normal "background" rate for extinction—that is, the normal extinction rate outside of times of mass extinction—is between one and four species per year.[33] The problem, however—as Raup himself has been quick to point out—is that paleontologists have very little confidence in that number for a variety of reasons, mostly related to sampling. The fossil record is overwhelmingly biased toward marine invertebrates with easily fossilizable hard parts (e.g., shells), which almost certainly means that the record of large vertebrates, insects, plants, and other organisms that are rarely preserved is drastically underrepresented. And it is precisely these kinds of organisms—for which the fossil record is incredibly poor—that are most affected in the current crisis.

A related problem is that, as discussed in chapter 5, paleontological studies of mass extinctions tend to focus on the extinction of higher taxonomic units—genera and families—for the simple reason that fossil data at the resolution of individual species is extremely spotty. There are good reasons for justifying some extrapolation from these higher taxonomic levels to the species level, but always with the proviso that exact calculations of species extinctions during mass extinctions—or, for that matter, during "background" times—are essentially impossible. What this means is that, in order to more faithfully compare geological and current extinction rates—to compare apples with apples, in other words—scientists should really estimate the current extinction rate not of species, but of genera or families. Unfortunately, as Raup put it to me, "This would be virtually impossible because we have almost no record of [current] extinction of higher taxa, [since] it doesn't happen often enough to be observable on human time scales" (Raup 2013). So, while we can be fairly certain that we are currently experiencing higher than normal rates of species extinction, we really have no idea— within perhaps several orders of magnitude—whether the current rate approaches those seen during events like the Cretaceous-Tertiary extinction, since (a) we don't have reliable estimates of species extinctions during those past events, and (b) we have no idea what the "normal" rate of species extinction is for the kinds of localized, endemic terrestrial invertebrates being affected today. Combine this with the fact that, as discussed above, even *current* species extinction rate estimates (e.g., the "seventy-four species per day" claim) may be off by orders of magnitude, and one begins to perceive the scope of the problem.

I feel compelled, once again, to point out that the difficulties involved in estimating rates of extinction in no way invalidate the concerns of biodiversity proponents or suggest that we are not experiencing a crisis of some kind. But is it a mass extinction? My sense is that, these days, most paleontologists would cautiously allow that yes, we either are experiencing or are in grave danger of triggering a mass extinction of terrestrial invertebrates, plants, and some vertebrate species (e.g., marine mammals, terrestrial megafauna, fish, and birds). Whether this is a "sixth extinction," though, is quite a bit more problematic. Mass extinctions have occurred many times in the history of life; while there is currently debate about the exact number (owing to disagreements about data interpretation and to conflicting definitions for what constitutes a "mass extinction"), life may have experienced at least twenty major extinction events over the past half-billion years.[34] The so-called "big five" extinctions are simply the most spectacular of these events, and there is still considerable paleontological debate about their causes, duration, and magnitudes.[35] We simply have little way of knowing whether the biodiversity crisis will reach the proportions experienced in these past events, and sadly it will only be possible to see the true picture thousands or millions of years from now, when it may be far too late for our own species. That fact in itself militates against inaction.

On the other hand, claims that we are experiencing a sixth mass extinction are not "merely" rhetorical; the analogy between past and present mass extinctions is meant to highlight the serious ecological and evolutionary consequences of extinction events—a subject that has received significant attention from paleontologists since the 1980s. Norman Myers, in particular, was an important early proponent of this analogy, and in several articles beginning in the mid-1980s he drew attention to how the work of paleontologists might influence the way we understand the current biodiversity crisis. In an essay published in the magazine *Natural History* in 1985, Myers was one of the first biologists to suggest that the current crisis might have serious, lasting evolutionary consequences. He compared the current depletion of biodiversity with extinction events of the past, noting that the present crisis deviated from earlier events in that species losses were taking place in a time frame of decades, rather than the millions of years observed in the geological past.[36] He also argued, drawing on recent paleontological studies of extinction and diversification by Jablonski and others, that the steep losses in tropical regions were especially worrying, since paleontologists had suggested that the tropics are vital sources for the emergence of new species—"powerhouses," as Myers put it—on which evolution-ary processes depend. Ultimately, Myers argued, perhaps the most sig-nificant outcome of the biodiversity crisis would be not merely be the loss of existing diversity but "the hiatus in evolutionary processes that they will cause"—a loss of creative potential that could set evolution back tens of millions of years (Myers 1985, 6). Here he quoted a line from an influential 1980 anthology by Soulé and Brian Wilcox (which Myers would repeat in several future essays): "'Death in one thing, an end to birth is something else'" (Soulé and Wilcox 1980, 1–8). Myers predicted that when all was said and done, "the impending upheaval in evolution's course could rank as one of the greatest revolutions of pa-leontological time," perhaps even rivaling "the development of aerobic respiration, the emergence of flowering plants, and the arrival of limbed animals" (Myers 1985).

Five years later, Myers reiterated these arguments in a more pro-fessional journal, where he also expanded and updated the paleonto-logical basis for his claims. Noting that the comparison between past and present extinctions had "hardly been touched upon in the profes-sional literature" (a somewhat hyperbolic if generally accurate state-ment, at least from a technical perspective), Myers summarized studies by Jablonski, Raup, and Sepkoski on the selectivity and recovery dy-namics during mass extinctions, highlighting the long-term evolution-ary consequences of extinction events.[37] Here he considered both the factors that appeared to have enabled species to survive extinction events, and the effect of mass extinctions on "the subsequent course of evolution," echoing Raup's and Jablonski's conclusion that while mass extinctions had generally been "selective" (in that differential survival can be correlated with adaptive features of species), they were "not necessarily 'constructive' in a Darwinian sense" (meaning that they did not "reward" adaptations which had been successful in the past, or pro-mote enhanced fitness in survivors; Myers 1990, 178). The moral of this story for Myers was that the evolutionary consequences of mass extinc-tion are unpredictable, and that paleontological studies suggest that survivors tend to "contain a disproportionate number of opportunistic species"—or, in other words, that "our descendants could shortly find themselves living in a world with a 'pest and weed' ecology" (Myers 1990, 180–81). Repeating the dire predictions of his earlier essay, Myers concluded: "We—or rather, our direct descendants—may well find that many evolutionary developments that have persisted throughout the Phanerozoic could be suspended if not terminated" (Myers 1990, 183).

What did paleontologists themselves think of these arguments? Myers was an important intermediary between the paleontological and conservation communities, since his engagement with paleontologists and their literature was much more serious and sustained than had been those of his colleagues Wilson, Ehrlich, Raven, and others. In his journal articles, Myers thanked Jablonski, Raup, and Sepkoski variously for "numerous discussions" and "many illuminating discussions over the years," and his summaries of paleontological evidence were generally careful and accurate. Myers's discussions of the evolutionary consequences of mass extinction were extremely influential in shaping the early biodiversity crisis discourse—especially for Wilson, who regularly turned to Myers's articles to substantiate empirical claims. This influence was a two-way street: Jablonski, who was a graduate student at Yale in the late 1970s when *The Sinking Ark* was published, recalls that Myers's environmental warnings were a significant inspiration for his own early studies of extinction in the geological past, and paleontologists including Jablonski and Sepkoski cited Myers appreciatively in their own articles on the relationship between past and present extinctions.[38]

At the same time, paleontologists tended to take a more conservative view than Myers and other biodiversity proponents toward drawing lessons from the past. Raup, who was an important early source of information for conservationists, later had something of a falling-out with Wilson and others over the use of his estimates of background rates for projecting current biodiversity loss as a "mass extinction." Recalling his involvement many years later, Raup remembered reading portions of the manuscript of Wilson's *Diversity of Life* and debating Wilson's "use of extinction rate estimates from the fossil record to evaluate present-day extinctions and the possibility of a 'sixth' extinction" (Raup, personal communication 2013). He and Wilson maintained a cordial relationship but, he said, "Soon, other people got involved in the correspondence, probably including [the prominent ecologist] Stuart Pimm, and I felt most uncomfortable. They wanted me to back off and I guess I did."

No record of this exchange survives in Wilson's correspondence (though Wilson thanked Raup for commenting on his *Diversity of Life* manuscript), but Wilson and Raup did exchange letters regarding some of Wilson's earlier publications. For example, in his comments on an article Wilson was preparing for a meeting at the American Philosophical Society in 1986, Raup raised several concerns related to comparing past and present extinction data. In the first place, he suggested, "I suspect that your estimate that the loss rate now is several orders of magnitude greater than is typical for geologic rates is a bit high." He added that the fossil record was "highly right-skewed" due to a phenomenon known as the "pull of the Recent" (Raup to Wilson June 5, 1986).[39] What this meant is that because of biasing factors like the volume of fossil-bearing sediment and the exposure of outcrops, more recent geological periods tend to be much better represented in the fossil record than earlier ones are, thus contributing to the misleading appearance of greater taxonomic diversity and longer species durations in the more recent eras of the history of life (and hence appearing as an uptick of diversity on the right-hand side of a diversification graph). Extinction rates may also appear to be artificially higher in earlier periods, since there is a much greater likelihood that representative fossils would either not be preserved or not be discovered, potentially truncating the survivorship durations of particular groups. As a related point, Raup noted, "We can rarely, if ever, work with small (and presumably short-lived) endemic species," by which he meant that "estimates of species longevity that come from the fossil record are probably high by one or more orders of magnitude." In other words, average extinction rates for geological time are probably much too low, since many species come and go without ever leaving a trace in the fossil record.

In his reply, Wilson acknowledged that Raup's remarks were "are all on target . . . they emphasize the key difficulty of which we are aware: the birth and death of species is one of the least worked and most important subjects of biology" (Wilson to Raup, June 10, 1986).[40] Nonetheless, Wilson defended his estimates for the simple reason that actual extinction rates were unknown. As he put it to Raup: "We don't know whether species 'hidden' in the geological record, that is, rare, local, or sibling, have different longevities from the ones recovered. If they do, what you say is correct. If not, my very crude estimate is defensible. I wouldn't put bets on either side." Despite that disclaimer, this issue turned up again a few years later, when Raup commented on Wilson and Ehrlich's article for the special "Biodiversity" issue of *Science*. Again Raup cautioned against reading too much into the appearance of higher diversity levels in the present: "This is very difficult to prove because the Pull of the Recent exerts such a strong bias in the same direction." He also argued that comparative diversity estimates based on rainforest biotas were especially tricky (Raup to Wilson, September 8, 1990).[41] Here Raup pointed to recent studies suggesting that "extensive tropical rain forests are geologically unusual," which presented an additional complication: if the current global biota was not representative of the average distribution of species in geological terms, then making comparisons between past and present could be highly misleading. As Raup put it, "Because so much biodiversity is tied up in the rain forests, global diversity may have fluctuated rather wildly." Finally, Raup reported on "some new analyses of Jack Sepkoski's data [that] show that the (five) big mass extinctions are simply the tail of an asymmetrical distribution of extinction intensities." Raup explained that he had come to believe that extinctions are generally clumped fairly closely in time, and that "typical time intervals up to about 100,000 years experience essentially no extinction. Thus, the mean rates of extinction we observe in the fossil record are made up of a lot of widely-spaced events of non-zero extinction rate—with the mass extinctions merely being the rarest and most intense." In other words, there might effectively be no such thing as "background" extinction, since Raup suspected that *all* extinctions took place in coordinated bursts, some simply larger than others.

Raup's criticisms did not suggest that he discounted the effects of human activity on current species extinctions. They did reveal, however, that Wilson and other colleagues were presented with information that complicated the tidy "sixth extinction" analogy, which for reasons best known to themselves they had chosen to essentially ignore, particularly in popular writings. By the early 1990s, paleontologists themselves began to actively take part in discussions about the biodiversity crisis, and in general their appeals struck a balance between grim warnings about the lessons of past mass extinctions and cautious disclaimers about the need for better understanding of biodiversity crises in both the past and the present. Raup himself framed his own 1991 book *Extinction: Bad Genes or Bad Luck?* (discussed in the previous chapter) with the observation that his subject bore on "the contemporary problems of endangered species, losses of biodiversity, and extinctions caused by human activities." He remarked, "The history of species extinction provides valuable perspective on global ecology of the present and the future" (Raup 1991, xi–xii). However, he made little further reference to the current crisis in the book, focusing instead on the evolutionary dynamics of mass extinctions in the past.

At the other end of the spectrum, in that same year the AMNH paleontologist Niles Eldredge published an urgent appeal provocatively titled *The Miner's Canary*, which combined a survey of mass extinction research with warnings about the potential future impact of biodiversity loss. In the opening of the book, Eldredge addressed the issue in terms that neatly summarized some of the central links between paleontology and biodiversity conservation:

> I have come to realize that these two separate threads—the remote past and the modern world—really have much to reveal to one another. They are simply strands of the same rope. Extinction—truly massive, global extinction—is indeed a fact of the history of life. Thanks especially to the demise of the dinosaurs . . . the public at large is at least passingly familiar with the idea of mass extinction. We are also more or less aware that species are disappearing at an alarming rate right now: Extinction is a fact of life in the modern world (Eldredge 1991, xvii).

Eldredge went on to define extinction in precisely the same way as pioneering paleontologists like Raup and Sepkoski—as "the loss of biological diversity, that is, the number of species"—and he argued for the need for "a general theory of extinction that relates past to present, and perhaps helps us see a bit more clearly the nature of our own present-day situation" (Eldredge 1991, xviii). As he developed the outlines for such a general theory, Eldredge drew heavily on the work of Raup and Jablonski, stressing both the role that mass extinctions had played in opening new evolutionary opportunities (its "creative" role), and also the tendency for extinction to "inhibit further evolution, as it removes genetic information that is essential for future evolution to occur" (Eldredge 1991, 8).

On the surface, Eldredge's argument would seem to support claims made by Myers and others that biodiversity loss presented a threat to the future course of evolution and, because of our dependency on global ecosystems, to the survival of the human species. But here Eldredge introduced a wrinkle, which he associated with a misunderstanding of evolution tied to a "Darwinian" view: "the mistake of thinking of evolution as a good thing." He explained that evolution itself is essentially neutral; it offers no guarantee of progress or improvement, but is simply the process that "has given us life's history (with a major role played by extinction); it will give us life's future, whatever form that happens to take." The fallacy, in other words, lay in thinking that in preserving biodiversity we were doing something noble for the future of the earth, or that we could escape anthropocentrism in addressing questions of human survival. The "ironic" fact, as Eldredge noted, was that "if we manage to survive . . . *that* will throw a damper on evolution more than will our or any species' extinction." He said our best course of action was simply to "conserve genetic diversity—ours and other species'—to maintain the status quo, and not because of some imagined effect this will have on the evolutionary future" (Eldredge 1991, 12).

In a sense, Eldredge was urging a practical anthropocentrism—"The bottom line is that the species that we must conserve is our own"—as an antidote to a more hubristic, philosophical anthropocentrism that endowed humans with a godlike power to stand outside of nature, controlling the future course of evolution. He may have been targeting claims like those advanced two years earlier by the environmentalist Bill McKibben, who in his bestselling *The End of Nature* argued that, through our industrial and agricultural footprint and technologies like nuclear weapons, humans had developed "the capacity to overmaster nature," and had "deprived nature of its independence" (McKibben 1989, 66, 48). It was true, Eldredge acknowledged, that "sedentary,

agriculture-based extinction is to the modern world what that asteroid . . . was to ecosystems at the end of the Cretaceous," but this hardly
meant that the human species now existed apart from nature (Eldredge
1991, 217). Rather, the increasing scope of human activities had only
underlined our inextricable relationship with the natural world:

> We have stepped beyond local ecosystems only to find ourselves as a part
> of the grand global ecosystem. We have not escaped nature—we only
> think we have, because we have stepped beyond the usual role of integra-
> tion into local ecosystems (Eldredge 1991, 220).

Instead of focusing on arguments based in the misguided belief that we
can transcend or control natural and evolutionary processes, we should
accept that we cannot help having an impact on the world around us,
and focus on the "purely selfish" goal of preserving "enough of the natural ecosystems intact so that the global system remains recognizably
what it has been throughout the history of our species so that we as a
species can survive" (Eldredge 1991, 221). Ultimately, Eldredge argued,
we can save ourselves only "by realizing that, though the rules have
changed, we will never escape intimate relations with the rest of the
biosphere" (Eldredge 1991, 229).

Biologists like Wilson and Myers seemed to realize that, to make the
important argument that biodiversity loss threatened to alter the course
of evolution permanently and irreversibly (thus elevating the crisis to
genuinely catastrophic proportions, on a par with thermonuclear war),
the cooperation of paleontologists was essential. Myers had been nag-
ging Wilson to highlight these evolutionary consequences since the
mid-1980s, arguing in a 1984 letter that "a basic impoverishment of
many processes of evolution" could be even more significant than the
"gross diminishment in the array of lifeforms on the planet," and urging him a year later to discuss the "impoverishing impact" of mass extinction after reading a draft of Wilson's major analysis of the crisis for
Bioscience (Myers to Wilson, October 11, 1984).[42] By 1991, Myers had
apparently grown frustrated that these consequences had been insuffi-
ciently promoted, and he lobbied Wilson (unsuccessfully) to intervene
with the editors of *Science* to include his own article on the subject in

the special issue on biodiversity planned for later that year. In his letter to Wilson, Myers confided that he had been discussing the possibility of coauthoring such a paper with Jablonski, but had learned that Jablonski would be contributing his own article to the issue.[43] Myers asked for the opportunity to "take a crack at it" himself, since he felt the topic was especially important "in light of the poo-pooing attitude of certain organization [*sic*] . . . primarily IUCN but increasingly WWF International too."

Three years later, Myers wrote to Wilson again, this time with a more urgent tone. Myers had persisted in his efforts to secure a prominent coauthor for an article on "the impact of the present biotic crisis on the future course of evolution (no less!)," with little success (Myers to Wilson, July 5, 1994).[44] He reported that both Ehrlich and Raven had claimed to be "too busy" to participate, and that Raup had declined because "he still doubts whether there is truly a biotic crisis underway." He confessed to Wilson:

> I am really at a loss on this one. . . . I think the paper should be written, since it will address an almost entirely neglected dimension of the biotic crisis. When I run a computer check in the library, I find not a single paper has tackled this mega-issue. Folks like Richard Dawkins and Steve Gould are completely retrospective in their approach. I am looking for somebody who will be pioneering and exploratory with me, even speculative (Myers to Wilson, July 5, 1994).[45]

Unfortunately for Myers, Wilson provided little assistance, so he next turned to Jablonski, whom he solicited for "a lengthy look at the notion of a joint paper for a *Science*-calibre publication" (Myers to Jablonski, July 12, 1994).[46] After outlining the proposed argument, Myers stressed the need for a coauthor "because I do not have the experience myself, nor the scientific insights and the professional experience—certainly not the palaeontological clout—to go it alone." What Myers had in mind, however, seems to have been something closer to an endorsement from Jablonski than to a genuine collaboration. Enclosing a manuscript with his letter, Myers explained: "What I have in mind—if you consider the enclosed draft is more or less on target—is that you expand it, re-jig it, re-orient parts of it if need be, and generally make it a paper to which you would be glad to append your name."

In the end, though, Myers was unsuccessful with Jablonski as well. He resorted to publishing a final, lengthy article on the topic for the journal *The Environmentalist*—a respectable publication, but hardly the "*Science*-calibre" journal he had hoped for. This essay largely reiterated the points he had made in his 1985 and 1990 articles on the same subject, even duplicating some of the language of the final paragraphs in the 1985 *Natural History* essay; but it also updated some of the discussion of the "longer-term consequences" of biodiversity loss, particularly drawing on Jablonski's recent studies of the importance of tropical environments as the "cradle" of diversification.[47]

Some of Jablonski's reluctance to coauthor with Myers may have been due to the fact that he was already in the process of publishing his own statement on the relationship between past and present extinction crises, which appeared in April of 1994 as an essay titled "Extinctions in the Fossil Record." Jablonski's article was adapted from a paper presented at "Estimating Extinction Rates," a 1993 symposium largely composed of biologists and ecologists, and sponsored by the Royal Society of London. It offered a brief overview of paleontological analysis of mass extinctions, with an eye toward discussing "some implications for today's biota" (Jablonski 1994, 11). Overall, Jablonski's message was cautiously optimistic about the application of paleontological studies to the present crisis, concluding that "the fossil record provides our only empirical data on what happens when biological communities collapse or disassemble, when increased extinction rates impinge on taxa of different relative vulnerabilities, when global warming or cooling occurs faster than species can adjust to local conditions, when ecological stresses ameliorate after prolonged or severe episodes, and so on" (Jablonski 1994, 15).

At the same time, however, Jablonski raised a number of concerns about comparisons between past and present rates, including a discussion of the inherent problems with paleontological data (as discussed above), the limitations of statistical analysis, and difficulties estimating species-level extinction from data for higher taxa. He acknowledged that the fossil record could "provide useful insights for conservation," but cautioned, "Palaeontological extinction data are extremely difficult to compare to present-day extinction rates" (Jablonski 1994, 13). In fact, he argued, rather than looking to paleontological analysis of mass extinctions, the more useful comparison might be with "palaeontological analyses of background extinction as a tool in assessing present-day extinction"—a suggestion he admitted might "be surprising." The key question was whether the current crisis really reflected the observed dynamics of mass extinctions; while Jablonski stressed, "I am not belittling the magnitude of today's problems," he nonetheless contended, "It is not clear that present-day disturbances, although extensive relative to the quietest times of Earth history, are on par with those that drive the major mass extinctions" (Jablonski 1994, 14).

In particular, Jablonski noted that whether or not the current biodiversity crisis equaled past mass extinctions in terms of an absolute number of species lost, the real issue was whether "the qualitative change in survivorship such as seen at the Cretaceous-Tertiary boundary has occurred today"—in other words, whether the durations of higher taxa had been artificially truncated with respect to their average durations over geological history. This was a claim for which he found "little evidence":

> So far as they are known, today's extinction patterns conform mainly to intensified versions of background expectations, with losses concentrated in endemic species and subspecies. The major mass extinctions operated on a different scale: genera endemic to single subcontinental provinces were lost preferentially, regardless of the geographic ranges of their constituent species (Jablonski 1994, 14).

Furthermore, and echoing Raup's private criticisms to Wilson nearly a decade earlier, Jablonski observed that it was often extremely difficult to neatly distinguish episodes of "mass extinction" from "background" rates in the fossil record, since "extinction magnitudes for the stratigraphic stages of the Phanerozoic form a continuous distribution," and "many impressive extinction pulses fail to stand significantly above background variance." In the end, what determines the impact of an extinction event is less a matter of *how many* species are removed than *which* groups are lost. As Jablonski put it, "Mass extinctions have such profound biological consequences because they bite deep into standing diversity and disrupt background selection regimes, not because they account for most species terminations" (Jablonski 1994, 14).

I recently asked Jablonski whether in the decades since he wrote this paper he has been given any reason to revise his assessment. Would he now acknowledge that we are in the midst of a "sixth extinction"? His response was careful and somewhat guarded, but it summarized the distance between the rhetoric of proponents of the sixth extinction concept and the paleontological analysis on which it is ostensibly based: "I would not say that we are in the middle of a mass extinction" (Jablonski interview, February 27, 2017).[48] Jablonski was neither challenging the reality of the biological diversity crisis nor arguing that paleontological analysis of extinction is not useful for understanding the current problem. Indeed, even in his 1994 article he refused to "deny the potential for long-term losses of similar scope and evolutionary impact to the major mass extinctions of the fossil record." Rather, Jablonski's point— similar to arguments made by Raup, Sepkoski, and other paleontologists over the past several decades—was that these kinds of comparisons can be made, and can only be useful, if the data they compare are commensurable—in other words, if we are comparing apples to apples.

Paleontologists are conservative by nature about making broad claims based on data from the fossil record because of its notorious "incompleteness," and I suspect that this conservatism rubs uncomfortably against the rather opposite tendency for conservation biologists to extrapolate wildly from equally incomplete information. This is perhaps natural: paleontologists, after all, deal with the deep past and are rarely called on to address current political problems, whereas conservationists, who deal with much shorter periods of time and with fickle politicians and public interest, must of necessity, as Soulé put it, "act before knowing all the facts." Indeed, over the past twenty years or so many paleontologists have deliberately moved toward advocacy positions regarding biodiversity, and in doing so they have reconsidered their roles as interpreters of the past. In his 1997 presidential address to the Paleontological Society, for example, Jack Sepkoski argued, "We can expand our role beyond informing other scientists and the public about life in the past [to] help inform scientists and the public about life in the future" (Sepkoski 1997, 537). As Sepkoski noted, paleontologists "are the only scientists who have ever seen biodiversity crises to their end, know consistent characteristics of species at risk, and have some idea of what happens in the aftermath" (Sepkoski 1997, 533).

Extinction, Diversity, and Culture

Are we, then, in the midst of a mass extinction? Does it matter what labels we use? From one perspective, to speak of a biodiversity "crisis" versus a "sixth mass extinction" is to make a distinction without a difference: as conservation biologists and paleontologists agree, we are experiencing a period of unusually high extinction rates, and even if we were somehow able to magically prevent even one more species from becoming extinct beginning tomorrow, the consequences for global ecology will still be profound in the future. Moreover, human beings bear the brunt of responsibility for this crisis, though we can and do debate whether this extinction event should be dated to the beginning of the Industrial Revolution of the early nineteenth century, to the advent of agriculture ten thousand years ago, or all the way back to the Pleistocene extinctions of Ice Age megafauna that may have been triggered by our remote ancestors more than 120,000 years in the distant past. In any event, as Eldredge put it in *The Miner's Canary*, for late 20th and 21st late-twentieth- and twenty-first-century society, "extinction is a fact of life in the modern world," and one that, like anthropogenic climate change, may make the lives of our children and grandchildren considerably more difficult than our own, despite our best efforts today.

On the other hand, some would argue that when scientific debates become part of political discourse — and vice versa, since it is never possible to cleanly separate these cultural domains—it is important that the terms we use are precise, that our data analysis is accurate, and that our projections are not willfully exaggerated. In conversation with me several years ago (before his death in 2015 at the age of eighty-two),

Dave Raup reacted strongly to my suggestion that "a little exaggeration" about biodiversity was OK if it motivated the public and politicians to take action. He said:

> Here I object strenuously. Global climate people are in serious hot water because they have been caught out so often in exaggerations. To be sure, some of the charges are merely the result of extreme advocates on the other side, but I think enough of them are real to have seriously degraded the climate change message. The same sort of thing has gotten evolutionary biologists in a lot of trouble for exaggerating the strength of the Darwinian model (Raup, personal communication 2013).

Raup was by his own admission a bit of a contrarian, so perhaps his analogy between biodiversity rhetoric and debates around global warming or creationism is slightly overblown. On the other hand, in comparison to the two cases he cited, biodiversity estimates are arguably significantly *more* fuzzy and uncertain than are climate models or evolutionary theory. These other topics certainly do not see the orders of magnitude of disagreement about basic facts that have been advanced in biodiversity discourse over just a few decades. Recall, for example, that in 1985 Wilson claimed, "The rate of extinction is now about 400 times that recorded through recent geological time." In just a decade that figure ballooned to "120,000 times above background," in Richard Leakey and Roger Lewin's popular book *The Sixth Extinction: Patterns of Life and the Future of Humankind* (Wilson 1985, 703; Leakey and Lewin 1995, 241).

It is important to note that these upward revisions are not due to any refinement of geological extinction rates or significant improvement in taxonomic knowledge of current species. Indeed, despite the efforts of people like Wilson to initiate a massive global biodiversity cataloging program, scientists have still identified only about 1.5 million species of eukaryotes in total (all organisms excluding bacteria). Rather, extinction figures have grown through a process of repetition and incremental modification of those very first estimates—by Terry Erwin and Wilson—that produced the "seventy-four species per day" number, with remarkably little additional empirical justification. Is it not reasonable to wonder whether a public mobilized by claims that we are losing as many as fifty thousand species per year might at least lose interest, if not faith in scientific expertise, if that number were revised significantly downward after we improve our taxonomic knowledge? What might happen if our "sixth extinction" turns out more closely to resemble a period of accelerated "background" extinction, as Raup and Jablonski have suggested? Especially in matters of science and technology, the public and politicians have become conditioned to reacting to events only when they achieve economies of scale that are "super-sized." To compete for attention in the news cycle, viral outbreaks must be incipient pandemics; earthquakes must register at the upper limits of the Richter scale; data leaks must be on the scale of petabytes. This mentality has often and rightly been connected both to the competitive attitude of late Cold War "big science" and to a state of twenty-first-century information overload in which so much of the "information" we consume is just noise. But I think the story presented in this book also plays a role. Western culture's addiction to superlatives is in part a product of a catastrophic mentality that has seen the scope of the projected apocalypse magnify exponentially in little more than a hundred years. At the turn of the twentieth century, the worry was that Western society was in decline; by the 1960s and 1970s it was that the human species might extinguish itself; and in our own day we hear about the potential for the "end of nature" and an irreversible curtailing of evolution itself.

But to go back to one of the major threads in this book, the cultural impact of this kind of "catastrophic thinking" has also manifested itself in some perhaps surprising ways, spilling into discussions far removed from nuclear war or endangered species, which have fundamentally affected the way Westerners have looked at their own societies. Here I am referring to the establishment of the notion that diversity itself has inherent value. One of the most dramatic cultural shifts documented in this book lies not in Western society's realization that the earth has undergone drastic physical and biological upheavals over its history, or that such "revolutions" challenge our sense of security in the unquestioned progress and survival of human beings. After all, Georges Cuvier presented his "catastrophic" geological theory at the very beginning of the nineteenth century, and an unsettling sense of pre- or even post-apocalyptic gloom hovered around many aspects of cultural discourse from the late nineteenth century onward. Rather, what is striking is how the new appreciation of extinction in the second half of the twentieth century contributed to a broader cultural and political reassessment of the value of diversity, both as a biological and a cultural "resource."

The establishment of the new valuation of diversity is particularly clear in the biological context, as we have seen in this chapter; but it has also translated directly to broader cultural associations with diversity, as has been noted by a number of observers. As David Takacs puts it,

> Some biologists who boldly assert that biodiversity is a normative good associate that claim with the more widely familiar one that cultural diversity is a normative good. As biologists link themselves with the forces promoting the multicultural ethic that has made normative and political headway in our society, different kinds of diversity thus become symbiotically and metaphysically linked in inherent "goodness" (Takacs 1996, 43).

I think that Takacs has this relationship exactly right. From the standpoint of the early twenty-first century, the inescapable conclusion is that, as he argues, "This thing called 'diversity' has been reified: a previously intangible or abstract concept has been made into a definable, graspable entity" (Takacs 1996, 45). Likewise, Timothy Farnham has asserted that the central value associated with "keep[ing] all our options open, preserving a greater variety of values by preserving the natural variety of the environment," has contributed directly to the belief "that with greater diversity—whether cultural or biological—comes greater value" (Farnham 2007, 7).

While it seems intuitively obvious from our present Western perspective that valuations of biological and cultural diversity are closely associated, it is extremely tricky to try to establish, in historical context, exactly how this relationship came to be. As Ursula Heise has observed, "Clearly, the cultural cachet that the concept of 'diversity' as accreted over the past half century in a variety of social spaces is hard to disentangle from scientific arguments" (Heise 2016, 30). The impossibility of fully disentangling such discursive threads has been a major theme of this book, and those attempting to determine the directionality of influences between explicitly biological and more broadly cultural valuations of diversity are confronted with a serious chicken-and-egg problem. Nonetheless, I will conclude this book with a perhaps bold claim that I feel is justified but probably deserves a book of its own to unpack its complexities and ramifications: A central component of late modernity's fascination with and valuation of diversity of all kinds—both biological and cultural—has been the establishment of the new extinction imaginary discussed in the second half of this book. That is to say, it is not the case that biodiversity rhetoric rode the coattails of a broader cultural diversity movement, as has been suggested by some commenters, including Farnham. Nor is it the case that the current discourse of diversity has taken its cues from values established in the biological context (though, as I will discuss below, there is some evidence for such a claim). Rather, this overall valuation of diversity is essentially a single phenomenon, and its underlying logic depends on the understanding of the threat extinction poses to the stability of complex systems, whether natural or humanmade.[49]

I want to be very clear; I am not arguing that this new extinction discourse is the *only* explanation for the popularity of values now associated with diversity of all kinds. There are a variety of sources and contexts that can explain the formation of twenty-first-century arguments for the normative value of cultural diversity, whether diversity is understood as ethnic, religious, linguistic, economic, ideological, or some other form of difference. This larger history of the idea of diversity—which deserves greater attention than it has received—naturally encompasses histories of race, economics, law, and politics stretching back at least two centuries. I am claiming, though, that the specific understanding of diversity as a phenomenon that contributes to the "health" of social or biological communities comes directly from—and would hardly make sense without—the essentially ecological perspective that has been the foundation of the modern extinction imaginary.

The effect of this shift can be seen most starkly in a comparison between Victorian and current attitudes held by Westerners about their ethical responsibilities toward their fellow human and nonhuman populations around the globe. As Paul and Anne Ehrlich put it in 1981,

> A little more than a century ago, many Westerners thought there was no need to behave ethically toward certain people because, as slaves or members of "inferior" races, they were excluded from the in-group. Today few Westerners—indeed, few people in any culture—would espouse such a view (Ehrlich 1986, 50).

This statement conjures Charles Lyell's claim that "if we wield the sword of extermination as we advance, we have no reason to repine at the havoc committed," since it is only natural for the stronger group to exterminate the weaker (Lyell 1830–33, 156). The fact that we would no longer blithely condone, as Lyell did, "the extirpation of savage tribes of men by the advancing colony of some civilized nation" certainly testifies to changes in Western views about race, shifting political ideologies, and alterations to other cultural sensibilities that lie well beyond the scope of this book. But the *way* in which these altered values and beliefs are justified—the specific reasons given for claims that diversity has intrinsic value—are deeply informed by the way that extinction has come to be identified as a source of threat and anxiety. In fact, I would go so far as to claim that the emergence of the extinction imaginary that has characterized the past several decades has been accompanied, as a kind of essential corollary, by the invention of the concept of "diversity," at least as it is now understood in both biological and cultural contexts. Richard Leakey and Roger Lewin's statement about changing valuations of biological diversity, in their 1995 *The Sixth Extinction*, can apply equally to discussions of cultural diversity:

> These days, whenever ecologists talk about biological diversity, they usually feel obliged to justify its value. A quarter of a century ago, no such obligation was felt, for few people bothered to talk about diversity at all. The question of its value therefore did not arise. Earlier still, around the turn of the [twentieth] century, the value of diversity wasn't an issue either, but for different reasons (Leakey and Lewin 1995, 124).

One very specific context in which we can see this clearly is in efforts by the United Nations and other organizations to establish a framework for preserving and valuing cultural and "biocultural" diversity during the 1990s and 2000s. A decade following the 1992 establishment of the UN Convention on Biological Diversity, UNESCO (the UN agency dedicated to "Educational, Scientific and Cultural Organization") produced its "Universal Declaration on Cultural Diversity," which framed the preservation of global cultural diversity in precisely the same terms in which biological diversity had been presented: "The Declaration aims both to preserve cultural diversity as a living, and thus renewable[,] treasure that must not be perceived as being unchanging but as a process guaranteeing the survival of humanity" (UNESCO 2002). The declaration made an explicit analogy between biological and cultural diversity, stating in Article 1 that "as a source of exchange, innovation and creativity, cultural diversity is as necessary for humankind as biodiversity is for nature."

This sense that cultural and biological diversity are not merely similar, but are in fact manifestations of the same phenomenon can be seen in the emergence around the same time of a new concept: "biocultural diversity." The term appears to have been coined at a 1996 conference called "Endangered Languages, Endangered Knowledge, Endangered Environments," held in Berkeley and sponsored by UNESCO, the WWF, and the newly founded Terralingua foundation (devoted to the protection of endangered languages; Maffi 2001). Papers from the conference were published in a volume titled *On Biocultural Diversity*, and they reflect the ways in which key elements of the biological understanding of extinction and diversity influenced contemporary discussions of cultural and linguistic endangerment. As the conference organizer Luisa Maffi explained in her introduction to the volume, species, ecosystems, and cultural and linguistic groups were "facing comparable threats of radical diversity loss," amounting, especially in the case of languages, to "an extinction crisis" of "unprecedented" proportions (Maffi 2001). The strategies Maffi described for averting this crisis were consciously and explicitly drawn from biodiversity conservation efforts: "Issues of linguistic and cultural diversity conservation may be formulated in the same terms as for biodiversity conservation: as a matter of 'keeping options alive' and of preventing 'monocultures of the mind'" (Maffi 2001). One of the greatest perceived threats inherent in the biodiversity crisis is the homogenizing effect of extinction—the "depauperization" of complex ecosystems in favor of the ecology of "pests and weeds" discussed above. Maffi and others used this as a clear analogy for cultural and linguistic diversity loss.

There are other ways in which biocultural diversity arguments drew on the same extinction logic as the biodiversity movement. For example, in his contribution "On the Coevolution of Cultural, Linguistic, and Biological Diversity," the linguist Eric Smith characterized cultural diversity as "the variation in culturally heritable information and its distribution across cultural lineages." Smith drew an explicit analogy between cultural and genetic diversity that even proposed concepts of phylogenetic branching, "drift," and isolation as factors in linguistic diversification (Smith 2001, 96–97). Another contribution to the volume noted the similar effects that "colonizing cultures" had on the reduction of both cultural and biological diversity. Just as colonizing societies reduce biological diversity by replacing indigenous flora and fauna with fewer, high-yield imported species of plants and animals, so too does globalization reduce linguistic diversity by imposing languages on colonized peoples. In this way, "the destruction of biodiversity and linguistic diversity have the same cause" (Wollock 2001, 250–51). Similar analogies can be found elsewhere in the literature on biocultural diversity, particularly in relation to endangered languages. In their 2000 survey of language extinction titled *Vanishing Voices*, Daniel Nettle and Suzanne Romaine stressed the value of linguistic diversity conservation as more than just a quantitative metric: "If some horrific catastrophe wiped out all the languages of western Europe tomorrow, we would lose relatively little of the world's linguistic diversity," since a very small percentage of the world's languages are spoken in Europe. More significantly, most European languages are quite similar structurally, so their loss would not dramatically affect the diversity of the *kinds* of languages that exist. On the other hand, Nettle and Romaine argued, if a similar number of languages were lost in South America or Southeast Asia, "the loss would be far more significant, because the divergence between languages there runs much deeper" (Nettle and Romaine 2000,

33–34). This is precisely the same logic applied to biodiversity conservation arguments that framed the issue as loss of "information" versus an absolute number of species—and both cases highlight the prospect of greatest diversity loss in locations with high proportions of small, locally endemic populations.

The conflation of biological and cultural diversity—and of their intrinsic value—was nowhere more evident than in a UNESCO booklet published in 2003, titled *Sharing a World of Difference: The Earth's Linguistic, Cultural, and Biological Diversity*. The booklet defines biocultural diversity as "interlinkages between linguistic, cultural, and biological diversity," asserting that "the diversity of life on Earth is formed not only by the variety of plant and animal species and ecosystems found in nature (biodiversity), but also by the variety of cultures and languages in human society" (Skutnabb-Kangas et al. 2003, 9). Beginning with now familiar arguments for preserving biodiversity, such as the "unforeseen consequences" of damage to the "delicate relationships" in ecosystems, and the endangerment of the "potential for adaptation" by the reduction of genetic diversity, it goes on to argue that "diversity is the basic condition of the natural world" (Skutnabb-Kangas et al. 2003, 9–10). "However," the booklet continues, "diversity is not only a characteristic of the natural world. The idea of 'diversity of life' goes beyond biodiversity. It includes cultural and linguistic diversity found among human societies" (Skutnabb-Kangas et al. 2003, 18). This cultural component to diversity, say the authors, can be thought of "as the totality of the 'cultural and linguistic richness' present within the human species"— a quantity analogous to the species and genetic richness of the global biosphere. Moreover, the argument describes the world's six to seven thousand languages as "the total 'pool of ideas'" represented in human culture, all of which are threatened by a "linguistic and cultural extinction crisis" (Skutnabb-Kangas et al. 2003, 28–29). Ultimately, the view presented in this booklet—and by a large segment of the biocultural diversity literature—is more than merely analogical: "Biological diversity and linguistic diversity are not separate aspects of the diversity of life, but rather intimately related, and indeed, mutually supporting ones. . . . The extinction crises that are affecting these manifestations of the diversity of life may be converging also" (Skutnabb-Kangas et al. 2003, 35).

These examples drawn from the biocultural diversity literature are instructive about some of the ways that extinction discourse has penetrated postmillennial cultural discussions of diversity, but of course they do not explain all of the wider valuations of diversity we see around us. When universities encourage applications from "diverse" communities, or when employers defend their record of diversity in hiring, they are almost certainly not consciously thinking of the value attached to these practices in terms analogous to arguments for preserving biological diversity. Yet I would caution against therefore dismissing my broader claim about the importance of an extinction imaginary in shaping even our casual valuations of diversity. We may be unconscious, or even unaware, of the "ecological" basis for valuing diverse communities, but I suspect that if we deeply probed our own rationales for encouraging diversity, we would find that they do resolve to some version of the arguments advanced for protecting biological diversity. Why, after all, should it matter whether a school or a workplace is diverse? What benefit is imagined to follow? If one were to respond that it is a matter of ethics or fairness to allow equal opportunity, I think that would be only a partial answer. Initiatives like the rules in the United States and Europe supporting affirmative action for women and particular minority populations are not, in the final analysis, about ensuring diversity, though they may contribute to this effect. Rather, they are designed to redress certain historical inequalities for specific groups. An employer would, in theory, be contributing to affirmative action by hiring a workforce composed of 80 percent women or 50 percent African Americans, which would have no effect toward creating a quantitatively diverse environment. This has indeed been a favorite argument of those who oppose affirmative action and similar programs on ideological grounds. From this perspective, such policies can actually *reduce* diversity by penalizing certain ethnic groups (especially noted in the case of Asians), or by limiting other kinds of diversity, such as political ideology.[50]

No; whether we acknowledge it or not, the implicit rationale for promoting cultural diversity in our workplaces, neighborhoods, schools, and the like is that exposure to a plurality of ideas, experiences, beliefs, backgrounds, traditions, abilities, and advantages will produce "healthier," more balanced, and more readily prepared individuals and societies. The loss of groups that make up this diverse cultural landscape is seen as being akin to the "extinction" of resources that can contribute to the resilience of our society—to its adaptability and resistance to sudden change. To be sure, there are many people who oppose this rationale and reject cultural diversity as "identity politics." But to encounter some of the bitterest complaints from that side of the political spectrum—which often argue that businesses and schools should be compelled to promote diversity of political affiliation, religious belief, and the like (generally a code for favoring white, heterosexual, politically conservative men)—is to get a sense of just how deeply, if inconsistently, these values have penetrated Western beliefs. If our society still has a very long way to go in grappling with what diversity really *means*, we seem remarkably certain *that* it is something that is good for us, and we fear its loss.

EPILOGUE: EXTINCTION IN
THE ANTHROPOCENE

Imagine this scenario: Over a short period of time, huge amounts of carbon dioxide, methane, sulfur dioxide, hydrogen chloride, and other gases are released into the atmosphere. The rapid dispersal of these gases has several significant effects, including the elevation of global temperatures by several degrees Celsius due to a "greenhouse effect," and the depletion of the protective ozone layer as the result of the inter-action of methane in the atmosphere, which reacts chemically with hydroxyl and suppresses the production of ozone. Terrestrial ecosys-tems are bathed in harmful ultraviolet radiation, and as temperatures rise above 35°C photosynthesis becomes significantly less efficient for plants and green algae. This creates an oxygen-depleted (hypoxic) envi-ronment, and severely impacts the diversity of vegetation and the ani-mals who depend on it, triggering cascading extinctions. At the same time, changes to the composition of the oceans are even more dramatic. Higher atmospheric temperatures raise the surface temperature of the oceans, producing a devastating effect on shallow-water ecological communities. Coral reefs die, and the complex ecosystems they sup-port are fatally disrupted, driving countless species of fish, marine in-vertebrates, and microorganisms to the brink of extinction. Outside the tropics where reef communities are generally found, increase in ocean temperatures shrinks the habitats of cold-water organisms by as much as 90 percent, leaving many species with simply nowhere to go.

Nor are the oceans immune to the effects of greenhouse gasses; absorption of CO_2, an acidic compound, along with even more toxic compounds like hydrogen sulfide raises the global pH, producing lethally acidified oceans. More devastating still is the pervasive deoxygnation of the seas, the result of a complex ecological and environmental cascade. In the first instance, oxygen becomes less soluble in warmer water, meaning that marine animals will find it harder to breathe. Secondarily, warmer temperatures promote increased organic decay, a chemical process that uses up oxygen. The oceans' phytoplankton (microscopic organisms that photosynthesize on the ocean surface) produce up to 85 percent of the world's oxygen, but photosynthesizing phytoplankton like green algae cannot survive in temperatures much above 32°C, nor can the zooplankton (protozoans such as radiolarians and foraminiferans) that feed on ocean bacteria. Zooplankton have an important role in maintaining marine oxygen levels, since they clear the ocean's surface of decaying algae and transfer organic matter to the sea floor, in the form of microscopic fecal pellets. As green algae and zooplankton die, they are replaced by opportunistic cyanobacteria, which have much higher temperature tolerances than phytoplankton such as radiolarians. Cyanobacteria—also known as blue-green algae, though strictly speaking they are not true algae—are among the oldest organisms on earth, and their photosynthesizing is thought to be responsible for the so-called Great Oxygenation Event that paved the way for multicellular life some 2.5 billion years ago. Unfortunately, cyanobacteria also produce a variety of toxins harmful to other marine life, and "algal blooms" are considered significant threats to marine ecosystems. Perhaps more significantly, zooplankton are unable to digest cyanobacteria (and, in any event, are likely killed by rising temperatures), meaning that the ocean's surface becomes choked with decaying organic matter, accelerating the consumption of marine oxygen. In a relatively short time, the world's oceans become anoxic, or nearly devoid of oxygen, and unable to support life.

The result of these interlinked, cascading ecological disturbances is mass extinction of a truly global scope. Virtually no group of organisms is spared: rising temperatures and reduced photosynthetic efficiency produces massive deforestation, hitting larger plants and trees especially hard. Whole orders of insects and other invertebrates that depend on the vanishing flora are wiped out, as are herbivorous vertebrates and the carnivores who prey on them. In the oceans the losses are even more widespread, and marine invertebrates with heavily calcified skeletons or who rely on those organisms for food or ecosystem services (including many groups of echinoderms, mollusks, arthropods, and microscopic zooplankton) suffer catastrophic extinctions. And both on land and in the water, larger animals are nearly lost; deprived of food and habitat, and especially sensitive to oxygen depletion because of their greater metabolic requirements, a great many vertebrates are simply unable to cope. In the course of perhaps only a few thousand years, levels of species extinction dwarf even the catastrophic event that killed the dinosaurs: as many as 70 percent of terrestrial vertebrates are gone, while marine species of all kinds suffer a mind-boggling loss of perhaps 96 percent of their diversity. The resulting earth is one nearly depleted of all life, and teetering on the brink of total extinction.

If you have paid any attention to the literature and media coverage of the current crisis of global warming, you might assume I am describing one of the more dismal projections produced by climate and ecological modeling for the coming century or two. But the scenario I have just sketched is not a prognostication for our near future, but an account of the deep past. It is a reconstruction of what likely happened more than 250 million years ago during what paleontologists consider the greatest mass extinction of all time, at the end of the Permian period.[1] The culprit in this instance is thought to have been the massive release of gases from a major system of "supervolcanoes" that erupted in parts of what are now Siberia and China, generally known as the "Siberian Traps" (a "trap" is the geologic term for the rock formed by lava flows). The end-Permian mass extinction is distinguished from the other members of the "Big Five" not only by its scope but by its suddenness: a recent study suggests that the majority of extinctions may have taken place over just a few *centuries*—not even a blink of an eye in geological terms.[2]

The parallels between the proposed end-Permian extinction scenario and our own worst fears about the impacts of global warming and biodiversity loss are inescapable, as many have observed. As the geologist Lee Kump describes it in a recent commentary in the journal

Science, "Voluminous emissions of carbon dioxide to the atmosphere, rapid global warming, and a decline in biodiversity—the storyline is modern, but the setting is ancient." Citing the event as a possible "analog for our future," Kump warns, "Our modern-day 'Siberian Trap' is fossil fuel burning, which is driving up atmospheric carbon dioxide to concentrations that Earth has not witnessed for millions of years" (Kump 2018, 113–14). The authors of the study on which Kump's comments are based—an analysis of the role of hypoxia in Permian extinctions—agree with this assessment; they conclude that global warming and oxygen loss could "largely account" for the end-Permian mass extinction, and forecast that their study "highlights the future extinction risk arising from a depletion of the world's aerobic capacity that is already underway" (Penn et al. 2018, 1130). In comments to the science journalist Carl Zimmer, one of the study's authors, Curtis Deutsch, is even more blunt, remarking that current global climate change "is solidly in the category of a catastrophic extinction event," and warning, "Left unchecked, climate warming is putting our future on the same scale as some of the worst events in geological history" (Shen et al. 2018, 205–23).

One of the central arguments of this book has been that, to paraphrase the wonderful 1984 editorial by Ellen Goodman discussed in chapter 5, every era gets the extinction story it "deserves." The optimistic Victorians saw extinction as a gradual succession of ever fitter species, leading to inevitable "improvement" that fit their view of history as an essentially progressive narrative. The gloomy Modernists of the early twentieth century, in contrast, found parallels between equally inevitable cycles of rise and decline in both natural and human history. With the advent of nuclear weapons and an increasingly frightening Cold War political and environmental backdrop, scientists, politicians, and the public became fascinated by sudden, cataclysmic events capable of obliterating life in the flash of an atomic blast or an asteroid impact. As the Cold War gave way to late-twentieth-century globalization and the triumph of a neoliberal political and economic order, however, both the immediacy and the stakes of the impending catastrophe shifted. Rather than being viewed as a sudden event, catastrophe came to be seen as a slow-motion affair, and human agency became associated—by analogy, at least—with the geological forces that shaped life in the past in contributing to a modern-day Sixth Extinction.

To put it another way, the scientific theories, cultural metaphors, and future prognostications bound up in "catastrophic thinking" collectively contribute to an imaginary around extinction that informs both our sense of the past and our prospects for the future. If this is the case, then what does the imaginary of our current moment (as of this writing, in late 2019) signify about our hopes and fears, and about our responsibility for and relation to the natural world we inhabit? Are we currently living in a continuation of the catastrophic thinking associated with the late-twentieth-century anxiety over biodiversity depletion, as described in this book's final chapter, or have we entered some new phase? What does it mean that, as some have suggested, we have come to see humanity not as some unpredictable external agent bringing death from above—an asteroid—but rather as an implacable geological force capable of altering the basic conditions for life on earth from within? As the science writer Peter Brannen puts it, "Today humanity plays the role of that primeval Siberian supervolcano" (Brannen 2017).[3]

There is a major line of argument in both the scientific community and broader popular discourse that suggests we have indeed entered just such a new phase of understanding, broadly associated with the proposition that we are now living in a new geological age: the so-called Anthropocene epoch. Originally introduced in the early 2000s as a proposed alteration to the official geological time scale, the Anthropocene concept is based on the observation that the footprint of human activity—atmospheric CO_2, radioactivity from nuclear testing, waste from plastics and other manmade compounds, widespread species extinctions, and other evidence of human environmental impact—is so profound that it will appear millions of years from now as a signature in the stratigraphic record. The idea was first widely presented by the Nobel Prize–winning atmospheric chemist Paul Crutzen in a 2002 article in the journal *Nature* titled "The Geology of Mankind"; it has since gained widespread notoriety and appeal in discussions ranging from geology and environmentalism to the humanities, as a way of conceptualizing the physical and psychological consequences of the unprecedented impact our species has had on the planet.

As Crutzen explained in his original statement, the current designation for the geological age in which we are living—an epoch known as the Holocene, which began at the end of the last major period of glaciation some 11,500 years ago—is insufficient to capture what has been distinctive about the signature left by humans since the industrial revolution of the eighteenth century. Crutzen's proposal has been taken up by various professional bodies responsible for ratifying changes to the geological timescale, including the International Commission on Stratigraphy and the International Union of Geological Sciences, and the change has been endorsed by the Anthropocene Working Group of the ICS, although no formal action has yet been taken. As a matter of geology, the Anthropocene is a somewhat controversial notion; there have been debates about when the Anthropocene should begin, with some favoring a more recent threshold (the start of industrialization, or even the advent of nuclear testing), while others advocating an earlier start, such as the spread of dynastic empires in Europe, Asia, and the Americas some two to three thousand years ago, or even the advent of agricultural societies six thousand years earlier (in which case the Anthropocene would effectively replace the Holocene). Still others have complained that geologists ought not be in the business of projecting into the future, and that the decision should properly await whoever is around to observe actual geological effects centuries or millennia down the road—if our civilization and species survive that long.

As a matter of dating and stratigraphic nomenclature, these questions can and will be decided on empirical grounds, and there is a possibility that the matter will be resolved even before this book is published. But this is not the primary function of the Anthropocene concept in our current imaginary around extinction. The significance of the Anthropocene discussion is, rather, cultural, to the extent that it signifies—as Crutzen, Will Steffen, and John R. McNeill argued in an influential 2007 article—the recognition of "a profound shift in the relationship between humans and the rest of nature," in which "humankind will remain a major geological force for many millennia, maybe millions of years, to come" (Steffen et al. 2007, 614–21). In this perspective, the Anthropocene is not merely a proposal for renaming a geological epoch, but a recognition of a radical reorientation of the relationship that humans perceive between ourselves and the rest of nature. As Jedidiah Purdy explains in his critical analysis of the Anthropocene, *After Nature*: "To define the Anthropocene is to emphasize what we think is important in that relationship, [and to recognize that] the familiar divide between people and the natural world is no longer useful or accurate" (Purdy 2018, 2). Independently of whether it is accepted as a stratigraphic designation, the notion of the Anthropocene symbolizes a new state of awareness about the permanence of human intervention in the natural world, and it crystallizes a host of new and preexisting anxieties and ambitions relating to climate change, biodiversity preservation, geoengineering, biotechnology, human population expansion, environmental and economic justice, and the future of humankind on or even beyond the planet Earth.

In other words, while the legitimacy and details of the Anthropocene can be debated by geologists, its relevance as a cultural touchstone is indisputable. A search of book and article databases returns thousands of results containing the term just in titles, and at least three scholarly journals are currently devoted to discussions of the concept and its consequences. This is to say nothing of the proliferation of the term in nonacademic discourse in newspapers, magazines, websites, and other forms of media; a current Google search returns nearly five million hits. The question is not *that* the Anthropocene signifies something, but *what* it signifies, particularly in the context of the narrative about extinction and its values presented in this book. Here the picture becomes more complicated: in terms of its message for the continuation of human civilization and our role as stewards of the natural world, there are currently at least two often quite conflicting understandings of the Anthropocene, which effectively inform two competing extinction imaginaries. The first recognizes the significant threat posed by global climate change, biodiversity loss, and other environmental crises, but regards these threats as challenges that human ingenuity can and will meet with technological innovations—such as geoengineering our atmosphere to reduce harmful greenhouse gases—that can reverse much of the damage we've done and even improve the quality of life for everyone. The second, much darker view sees the Anthropocene as the terminal moment for humanity, the culmination of our collective hubris—which will result in a dramatic reversal of our dominance of the planet, if not the extinction of our species.

It's worth pointing out that these twin themes—cautious optimism and extreme pessimism—have been paired in the extinction imaginaries at other moments as well. They represent two fundamentally human reactions to crisis—panic and doomsaying on the one hand, and resilient hopefulness on the other—that can and do coexist, even in the mind of a single person. E. O. Wilson's message about biodiversity, after all, has been alternately deeply pessimistic and guardedly optimistic, and it is clear that the sometimes dire rhetoric he has used to characterize the biodiversity crisis has been calculated to spur people to action, not to encourage despair and apathy. There have been some genuinely pessimistic extinction discourses in the past—for example, Oswald Spengler's forecast for the "decline" of Western civilization as an inevitable outcome, or various predictions of a similarly inevitable nuclear Armageddon during the Cold War era. Generally speaking, though, catastrophic thinking has often been the prelude to constructive action. It may simply be human nature to freak out before taking a deep breath and attacking the problem. If we are so inclined, we might even posit some speculative evolutionary explanation for this panic-then-recover pattern to crisis response, perhaps from our early experience as a vulnerable species on the African savannah.

Having said this, however, it is striking how widely separated the extremes are in the reactions to the Anthropocene. They seem to present not just different phases in the psychological absorption of a crisis, but basically alternative visions of the future. The historian Gregg Mitman puts this disconnect very well in a recent analysis of Anthropocene discourse:

> The first charts an environmental future of the "good Anthropocene," where technoscience provides the innovative tools for fixing a warming planet. The second propels us to a more dystopic environmental future, or at least a future filled with uncertainty, loss, and mourning in the face of accelerating species extinction and a world increasingly divided by those who have the means to survive and those who do not (Mitman 2018, 59).

One of the ways that the extinction imaginary of the Anthropocene differs from some previous incarnations is in its inbuilt fatalism. As proponents of the Sixth Extinction concept have frequently stressed, we are *already* well into an extinction event, and the species we have lost are irreplaceably gone. We can try to slow the pace of extinction, but we cannot recover what we have lost (unless we consider some of the far-fetched proposals to "de-extinct" particular species using cloning techniques, which will be discussed presently). Likewise, one of the central claims of climate change activists is that we have already passed the inflection point at which certain aspects of climate are reversible; like Huck and Jim traveling down the Mississippi in *Huckleberry Finn*, we've passed the turnoff point and there's no going back—the river is going to carry us forward whether we like it or not. One of the distinguishing features of the Anthropocene—and an argument for considering it as being part of a distinctly new extinction imaginary—is that, whether or not one sees the situation as redeemable, the crisis is upon us, and is not left to the imagination of some future event. When D. H. Lawrence wrote in 1928, "The cataclysm has happened, we are among the ruins," or when Jean Baudrillard exclaimed in 1989, "The explosion has already occurred, the bomb is only a metaphor now," the sense of living postapocalypse was nonetheless a metaphor. Global warming and biodiversity loss do not signify some other imagined catastrophe, or at least they do not *only* do so; they *are* the catastrophe, and they have most definitely already happened. There is nothing metaphorical about thousands of annual species extinctions or melting icecaps.

How we react to the experience of living through a catastrophe—whether we regard it as an apocalypse or just a challenge to be met—is another matter, though, and the response shapes very different visions of the future. From the very start, the optimistic view was baked into the notion of the Anthropocene itself. In his foundational 2000 article, Crutzen concluded that the challenges of climate change "will require appropriate human behavior at all scales, and may well involve internationally accepted, large-scale geo-engineering projects, for instance to 'optimize' climate" (Crutzen 2002, 23). Likewise, in their 2007 article Crutzen et al. suggested that "drastic options," such as sequestering CO_2

and dispersing aerosols into the atmosphere to block sunlight, may be required, though they did acknowledge that this strategy is "a highly controversial topic" (Steffen et al. 2007, 619).

In a similar vein, political scientist Amy Lynn Fletcher has proposed that pursuing technoscientific fixes for environmental and ecological damage caused by humans offers a redemptive counternarrative to the pessimistic gloom and mistrust of science often found in environmental critique. In her book *Mendel's Ark: Biotechnology and the Future of Extinction*, Fletcher writes: "To propose that biotechnology may someday allow us to undo the environmental damage we've done is to move from the twentieth century's environmental rhetoric of crisis to a new rhetoric of hope, to create a promissory wilderness which includes not only the species alive today but the multitude of species we thought irretrievably lost" (Fletcher 2014, 5). Specifically, Fletcher—a proponent of so-called "de-extinction," or the resurrection of extinct species through cloning and gene editing—argues that the use of biotechnology presents not just a practical fix to species loss (which she grants cannot possibly be fully recouped by technology), but a kind of pledge on the part of humanity that signals our optimistic spirit. Cloned extinct species are "promissory objects"—tokens of our commitment and ability to heal as well as harm—as much as they are concrete steps towards reversing biodiversity loss. While many of the de-extinction proposals focus on charismatic extinct species of little obvious ecological import, their symbolic value in the popular imagination is what matters most. She writes: "The idea of cloning a woolly mammoth is a socio-technical imaginary that embodies our fear of the present environmental crisis and our desire to save and create the future through biotechnological innovation" (Fletcher 2014, 91).

Indeed, an explicit rejection of apocalyptic thinking is a central feature of many commentaries on what Mitman called the "good Anthropocene." In one of the most passionate examples of this genre, the book *The Anthropocene: The Human Era and How It Shapes Our Planet*, the German science journalist Christian Schwägerl argues, "The Anthropocene is an anti-Apocalyptic idea, par excellence; an 'Apocalypse No' instead of an 'Apocalypse Now'" (Schwägerl 2014, 72). Schwägerl, who has collaborated with a group of historians and cultural critics in Berlin who have been prominently associated with the technological interventionist Anthropocene discourse, insists not only that climate change is remediable, but more importantly that the Anthropocene presents an opportunity to collectively reimagine the human future in ways that will "help people see themselves as active, integrated participants in an emerging new nature that will make the earth more humanist rather than just *humanized*" (Schwägerl, 2014, 33). In other words, while the current ecological crisis does indeed seem "frightening," it has provided an impetus to break free of the "narcissistic" tendency to regard ourselves as being set apart from the rest of nature. In this way, Schwägerl contends,

> If we take the Anthropocene idea seriously, it can help shape our present behavior in a positive way. Rather than defining humanity as the destroyer of nature, the Anthropocene casts people in an affirmative, long-term role. It is neither about facing an ecological apocalypse, nor harkening back to the "good old days." The Anthropocene is not a ticking time bomb, nor is it an end-of-the-world scenario (Schwägerl 2014, 72–73).

On the opposite end of the spectrum are commentators who have explicitly embraced apocalyptic thinking in their characterization of climate change and the Anthropocene. If the title of Roy Scranton's 2015 book *Learning to Die in the Anthropocene* doesn't make his point clearly enough, the opening sentence leaves nothing to the imagination: "We're fucked. The only questions are how soon and how badly" (Scranton 2016). Scranton—a literary scholar and journalist who draws prominently in his reflections from his deployment in the Iraq War in the early 2000s—rejects the notion that humanity can somehow come through its current crisis unscathed: "If *Homo sapiens* survives the next millennium, it will be survival in a world unrecognizably different than the one we have known for the last 200,000 years" (Scranton 2016). He also dismisses the notion that the Anthropocene is merely "the latest version of a hoary fable of annihilation" or an episode of mass "hysteria." Rather, he insists it is a "fact," and says, "We have likely already passed the point where we could have done anything about it." The only recourse humanity has left, he argues, is "to learn to die not as individuals, but as a civilization."

Surprisingly, however, Scranton's fatalism does not lead ultimately to complete pessimism, and here we may have a clue about whether there is, in fact, a coherent extinction imaginary that can be teased out of competing Anthropocene discourses. While we may have "failed to prevent unimaginable global warming," as he argues, so that our "global capitalist civilization as we know it is already over," Scranton nonetheless holds out a glimmer of hope: humanity yet has the possibility to "survive and adapt to the new world of the Anthropocene, if we accept human limits and transience as essential truths, and work to nurture the rarity and richness of our collective cultural heritage" (Scranton 2016). One consequence of this process is an essential reimagination of Western values and narratives—or, as Scranton puts it, "letting go of this particular way of life and its ideas of identity, freedom, success, and progress" (Scranton 2016). The book, then, concludes in an elegiac but cautiously optimistic tone:

> Wars begin and end. Empires rise and fall. Buildings collapse, books burn, servers break down, cities sink into the sea. Humanity can survive the demise of fossil-fuel civilization and it can survive whatever despotism or barbarism will arise in its ruins. We may even be able to survive in a greenhouse world. Perhaps our descendents [sic] will build new cities on the shores of the Arctic Sea, when the rest of the Earth is scorching deserts and steaming jungles. If being human is to mean anything at all in the Anthropocene, if we are going to refuse to let ourselves sink into the futility of life without memory, then we must not lose our few thousand years of hard-won knowledge, accumulated at great cost and against great odds. We must not abandon the memory of the dead.

Observing current scientific efforts to preserve archives of biological diversity, Scranton suggests that ultimately human civilization "must build arks: not just biological arks, to carry forward endangered genetic data, but also cultural arks, to carry forward endangered wisdom" (Scranton 2016).

If there is a coherent extinction imaginary present in the Anthropo-cene discussion, then, it is one that imagines ecological crisis as a pre-lude to a new chapter in human history. We might perhaps circle back and connect this view to the original meaning of apocalypse, as a cleans-ing fire that ushers in a new age or moral order for human society. There is something unmistakably biblical in this line of thought, despite the avowedly secular tone of most Anthropocene literature. Just as many apocalyptic religious sects have imagined that they were experienc-ing some final crisis predicted in scripture, Anthropocene proponents seem drawn to a similar moral drama, in which humanity is facing a test through which it will either emerge into a new, healthier era of pro-ductivity, stewardship, and humanistic values, or leave the stage for the benefit of what comes next. In other words, either we will learn to ap-preciate and protect the diversity of life that surrounds and sustains us as a species, or—like the trilobites, dinosaurs, and mastodons before us—we will become a statistic on some far-future graph of diversifica-tion and extinction, compiled by our hypothetical successors.

It is the outcome of that test is that remains uncertain, and this uncertainty contributes to the gap between optimistic and pessimis-tic readings of the Anthropocene. In some readings, environmental crisis is like the biblical story of the Garden of Eden, in which painful change will spur our species to new technological innovations or even a new start on other worlds. In others, it is like an angry God punish-ing humanity for its hubris and wickedness. There is something almost gleeful in the way some commenters imagine this negative outcome— as, for example, in David Wallace-Wells's recent book *The Uninhabit-able Earth*, which presents a catalog of horrors facing our species in the coming century, and paints a scenario of "a new kind of cascading vio-lence, waterfalls and avalanches of devastation, the planet pummeled again and again, with increasing intensity and in ways that build on each other and undermine our ability to respond . . . subverting the prom-ise that the world we have engineered and built for ourselves, out of nature, will also protect us against it, rather than conspiring with dis-aster against its makers" (Wallace-Wells 2019, 21). We might conclude that the current cultural fascination with apocalyptic entertainment— a seemingly inexhaustible appetite for stories of catastrophic plagues, natural disasters, and of course zombies—is the psychological projection of our own collective guilt and fatalism about a much more real and immediate catastrophe, both more complete and too prosaic to be translated to fiction.

If there is one Anthropocene scenario that seems almost too painful for either end of the spectrum to consider, it is a catastrophe with no final resolution, and with no moral message. The alternative to both a redemptive human transformation and an apocalyptic extinction is that, as Purdy has suggested, humanity merely continues to struggle along in misery. In an article appropriately titled "Anthropocene Fever," Purdy writes: "For all the talk of crisis that swirls around the Anthropocene, it is unlikely that a changing Earth will feel catastrophic or apocalyptic. . . . Indeed, the Anthropocene will be like today, only more so" (Purdy 2015). That is to say, the crisis will continue to disproportionately affect the part of the world that is already most miserable, and may largely spare the wealthiest and most developed societies, exacerbating already profound global disparities. As Purdy explains in *After Nature*, "The disasters of the Anthropocene in our near future will seem to confirm the rich countries' resilience, flexibility, entrepreneurial capacity, and that everlasting mark of being touched by the gods, good luck, [while] amplifying existing inequality" (Purdy 2018, 46). The result of this version of the Anthropocene, which Purdy labels "the *neoliberal* Anthropocene," is neither a glorious new society nor an earth purged of human interference, but a persistent dystopia of ever-widening inequality and economies of suffering.[4] Of all of the extinction imaginaries this book has considered, and of all the imagined outcomes these discourses have presented, this one strikes me as perhaps the most catastrophic and also the most plausible. In the end, it may be more palatable for Western culture to imagine its own complete extinction, or to conjure deus ex machina technological fantasies of utopian deliverance, than it is to conceive of an existence in which concrete, reasonable sacrifices are made by the fortunate of our species so that all human beings can experience a decent quality of life.

ACKNOWLEDGMENTS

A great many people have contributed to my thinking as it has evolved over the several years during which I have researched and written this book. I am especially indebted to my friends and colleagues at the Max Planck Institute for the History of Science in Berlin, where I was incredibly fortunate to spend six of the happiest years of my academic career. The book's origin goes back to an invitation in 2010 to participate in a "working group" there (a particular way of organizing collective intellectual projects at the institute) called "Endangerment and its Consequences," led by Fernando Vidal and Nelia Dias. I benefited immensely from workshops and discussions with members of that group, which were formative for this project. More broadly, the intellectual atmosphere provided by Department II of the Max Planck Institute — and the model and example set by its director, Lorraine Daston — were the ideal incubator for writing a book like this. At various stages of its development I had the opportunity to discuss this project in formal workshops and informal conversations with a great many of the smartest, most generous scholars in the world. I also had the enormous good fortune to be able to spend several years free of teaching and other obligations to research and write. My gratitude to Raine, to my colleagues, to visitors to the Max Planck Institute, and of course to the institute's wonderful library staff (who were able to provide any book or article I needed, no matter how obscure, at virtually a moment's notice) is boundless.

There are far too many people who provided helpful suggestions, insightful criticism, and sounding boards for me to thank everyone individually. Nonetheless, for inspiring conversation, comments on drafts of chapters or articles, or opportunities to speak or publish about various aspects of this project, I owe great thanks to Elena Aronova, Mark Barrow, Jenny Bangham, Etienne Benson, Josh Berson, Dan Bouk, Paul Brinkman, Mark Borrello, Lino Camprubi, John Carson, Jamie Cohen-Cole, Henry Cowles, Angela Creager, Helen Curry, Stephanie Dick, Sébastien Dutreuil, Sebastian Felten, Justin Garson, Michael Gordin, Oren Harman, David Jablonski, Boris Jardine, Judy Kaplan, Philipp Lehmann, Rebecca Lemov, Scott Lidgard, Erika Milam, Staffan Müller-Wille, Lynn Nyhart, Christine von Oertzen, Michael Ohl, Anya Plutynski, Sadiah Qureshi, Joanna Radin, Bob Richards, Lukas Rieppel, Sahotra Sarkar, Alistair Sponsel, Hallam Stevens, Marco Tamborini, Paul White, and Andrew Yang. My research assistants Julia Jägle (at the Max Planck Institute), and Lydia Crafts and Leanna Duncan (at the University of Illinois) have provided invaluable help in researching and preparing this book. I am also grateful for my current colleagues in the Department of History at the University of Illinois at Urbana-Champaign; a more welcoming and collegial group could not be hoped for.

Three people stand out as having been not only supporters of this project but important mentors for many years, and they deserve special recognition. Michael Ruse has encouraged and cajoled me for nearly two decades, and has often provided critical feedback, warm hospitality, and generous friendship at crucial moments. Martin Rudwick has inspired me for years with the example of his peerless scholarship and with his warm encouragement of my own work, including feedback and conversations about this book project. And the late David M. Raup, one of the true giants of twentieth-century paleontology, informed this book both directly and indirectly. Dave was probably the single most important theorist of the role of mass extinctions in the history of life, and I was fortunate to have him as a wise interlocutor and a friend. He offered critical observations and suggestions at the beginning of this project, and I dearly wish he was here to see the final product. He is greatly missed, and he is among those to whom this book is dedicated.

Special thanks are also due to the staff at the University of Chicago Press, and particularly to my editor, Karen Darling, whose enthusiasm and support for this project have been unflagging. Karen and the series editor, Adrian Johns, have provided much important encouragement and guidance along the way, from the project's initial conception to the finished book. Finally, my deepest and most important gratitude is to my family, who have always encouraged and sustained me. Thank you to my mother Maureen and my (late) father Jack; to Christine, Ella, and Sid; and of course to Teri, my best friend and first reader.

NOTES

Introduction

1. The official stratigraphic designation for what was for many years known as the early "Tertiary" period (T) is now officially recognized as the "Paleogene" (Pe or Pg). The term "Tertiary" is still widely used, however, to informally designate the geological period beginning roughly 66 million years ago, and it will be used thoughout this book.

2. E. O. Wilson, ed. *Biodiversity* (Washington: National Academy Press, 1988), 11–12.

3. On the topic of "imagination" as a way of understanding our cultural reaction to extinction, the literary scholar Ursula Heise's *Imagining Extinction: The Cultural Meanings of Endangered Species* (Chicago: University of Chicago Press, 2016) offers an insightful and evocative meditation on the way that "narratives" of extinction, especially in contemporary literature, have contributed to and been influenced by scientific discourse.

4. For interested readers, my use of the term "extinction imaginary" is similar to historian Sarah Maza's use of the term "social imaginary" in her study of revolutionary-era France, *The Myth of the French Bourgeoisie*. Maza defines the social imaginary as "the cultural elements from which we construct our understanding of the social world," including contemporary political and academic discourse, fiction, social commentary, and bureaucratic records. Sarah Maza, *The Myth of the French Bourgeoisie: An Essay on the Social Imaginary, 1750–1850* (Cambridge, MA: Harvard University Press, 2003), 10.

5. The complex interaction between scientific literature, popular and political writing and statements, artistic and literary representations, and indeed any "semiotic" web in which "signs" have particular meaning in relation to one another is often referred to as "discourse," another academic term I will sometimes use in this book — for example, in describing a shifting "extinction discourse." The term "extinction discourse" is central to the analysis of scientific and cultural values and beliefs surrounding extinction during the nineteenth century in the literary scholar Patrick

Brantlinger's book *Dark Vanishings: Discourse on the Extinction of Primitive Races, 1800–1930* (Ithaca, NY: Cornell University Press, 2013).

6. David Sepkoski, *Rereading the Fossil Record: The Growth of Paleobiology as an Evolutionary Discipline* (Chicago: University of Chicago Press, 2012).

7. David Raup to Thomas J. M. Schopf, January 28, 1979. Thomas J. M. Schopf papers, Smithsonian Institution archives.

Chapter 1

1. Arthur O. Lovejoy, *The Great Chain of Being: A Study of the History of an Idea* (New Brunswick, NJ: Transaction Publishers, 2009).

2. This essay, generally attributed to Linnaeus, first appeared in 1749 in Latin as a dissertation defended by Linnaeus's student Isaac Biberg. However, it was common practice at that time for dissertations to be dictated to the student by the professor and published under the student's name. My references to this text are to the 1762 English translation of the essay.

3. The most accessible introduction to nineteenth-century geological debates is Martin J. S. Rudwick, *Earth's Deep History: How It Was Discovered and Why It Matters* (Chicago: University of Chicago Press, 2014). For a deeper dive into this topic, Rudwick's twin volumes *Bursting the Limits of Time: The Reconstruction of Geohistory in an Age of Revolution* (Chicago: University of Chicago Press, 2005), and *Worlds before Adam: The Reconstruction of Geohistory in an Age of Reform* (Chicago: University of Chicago Press, 2008) are indispensable. Mark V. Barrow's *Nature's Ghosts: Confronting Extinction from the Age of Jefferson to the Age of Ecology* (Chicago: University of Chicago Press, 2009) also offers an excellent overview of the history of extinction, primarily from the perspective of North America.

4. François-Xavier Burtin, "Révolutions generals" (1789), quoted in Rudwick, *Bursting the Limits of Time*, 200.

5. Georges Cuvier, "A Discourse on the Revolutions of the Surface of the Globe," reproduced and translated in Martin J. S. Rudwick, *Georges Cuvier, Fossil Bones, and Geological Catastrophes: New Translations & Interpretations of the Primary Texts* (Chicago: University of Chicago Press, 1997), 190.

6. Jean Baptiste Pierre Antoine de Monet de Lamarck, *Zoological Philosophy: An Exposition with Regard to the Natural History of Animals* (Chicago: University of Chicago Press, 1984), ch. 7.

7. Lamarck, *Zoological Philosophy*, ch. 3.

8. Giambattista Brocchi, *Subapennine Fossil Conchology* (1814), partial trans. in Stefano Dominici, "Brocchi's Subapennine Fossil Conchology," *Evolution, Education, and Outreach* 3 (2010): 588.

9. Giamvattista Brocchi, *Mineralogical Treatise* (1807), trans. in Dominici, "Brocchi's Subapennine Fossil Conchology," 586.

10. Brocchi, *Subapennine Fossil Conchology*, in Dominici, "Brocchi's Subapennine Fossil Conchology," 592.

11. For a discussion of Brocchi's contribution to subsequent theories of organic change,

see Niles Eldredge, *Eternal Ephemera: Adaptation and the Origin of Species from the Nineteenth Century through Punctuated Equilibria and Beyond* (New York: Columbia University Press, 2015).

12. The "Scottish poet" Lyell refers to is Robert Burns, and the poem was Burns's 1785 "To A Mouse," which takes the voice of a farmer expressing regret for disturbing a mouse's home with his plow: "I'm truly sorry Man's dominion / Has broken Nature's social union, / An' justifies that ill opinion, / Which makes thee startle, / At me, thy poor, earth-born companion, / An' fellow-mortal!" Burns' poem also contains the famous couplets "The best laid schemes o' Mice an' Men / Gang aft agley,/ An' lea'e us nought but grief an' pain, / For promis'd joy!"

Chapter 2

1. In "Van Diemen's Land: Copies of All Correspondence between Lieutenant-Governor Arthur and His Majesty's Secretary of State for the Colonies, on the Subject of the Military Operations Lately Carried On against the Aboriginal Inhabitants of Van Diemen's Land" (1831), British parliamentary papers, 56.

2. "Papers Relative to the Condition and Treatment of the Native Inhabitants of Southern Africa, within the Colony of the Cape of Good Hope, or beyond the Frontier of That Colony. Part I. Hottentots and Bosjesmen; Caffres; Griquas" (1835), British parliamentary papers, 40 and 175.

3. Testimony of Archdeacon Broughton, "Report from the Select Committee on Aborigines (British Settlements); Together with the Minutes of Evidence, Appendix and Index" (1836), British parliamentary papers, 14–15.

4. The problems with the term "social Darwinism" as a historical category are nicely and concisely summarized by Diane Paul in her essay "Darwin, Social Darwinism, and Eugenics," in *The Cambridge Companion to Darwin*, edited by Jonathan Hodge and Gregory Radick (Cambridge: Cambridge University Press, 2003), 214–39.

5. For readers interested in a detailed history of eighteenth- and nineteenth-century European ideas about race, an excellent overview is provided in David N. Livingstone, *Adam's Ancestors: Race, Religion, and the Politics of Human Origins* (Baltimore: Johns Hopkins University Press, 2008).

6. See Martin S. Staum, *Labeling People: French Scholars on Society, Race, and Empire, 1815–1848* (Montreal: McGill–Queen's University Press, 2003).

7. Charles Darwin, "February 1835." DAR42.97–99. trans. and ed. John van Wyhe. Darwin Online, http://darwin-online.org.uk/.

8. See, e.g., Niles Eldredge, "Experimenting with Evolution: Darwin, the *Beagle*, and Evolution," *Evolution, Education, and Outreach* 2, no. 1 (2009); P. D. Brinkman, "Charles Darwin's Beagle Voyage, Fossil Vertebrate Succession, and 'The Gradual Birth & Death of Species,'" *Journal of the History of Biology* 43, no. 2 (2010): 363–99.

9. Charles Darwin, "Red Notebook," in *Charles Darwin's Notebooks, 1836–1844: Geology, Transmutation of Species, Metaphysical Enquiries* (Ithaca, NY: Cornell University Press, 1987), 133.

10. E.g., Robert J. Richards, "Darwin's Theory of Natural Selection and Its Moral Purpose," in Michael Ruse and Robert J. Richards, eds., *The Cambridge Companion to
the Origin of Species* (Cambridge: Cambridge University Press, 2009); Peter Bowler,
"Malthus, Darwin, and the Concept of Struggle," *Journal of the History of Ideas* 37
(1976): 631–50; Trevor Pearce, "'A Great Complication of Circumstances': Darwin
and the Economy of Nature," *Journal of the History of Biology* 43 (2010): 493–528.

11. Thomas Malthus, *Essay on the Principle of Population*, 6th edition (London: 1826), 3.

12. Charles Darwin, "Essay of 1842," in Francis Darwin, ed., *The Foundation of the Origin of Species* (Cambridge: Cambridge University Press, 1909), 51.

13. Charles Darwin, "Essay of 1844," in *Foundation of the Origin of Species*, 146.

14. E.g., David Takacs, *The Idea of Biodiversity: Philosophies of Paradise* (Baltimore:
Johns Hopkins University Press, 1996); Sahotra Sakar, *Biodiversity and Environ-
mental Philosophy* (Cambridge: Cambridge University Press, 2002); James Mac-
Laurin and Kim Sterelny, *What Is Biodiversity?* (Chicago: University of Chicago
Press, 2008).

15. Charles Darwin, *On the Origin of Species*, original first edition, e.g., 18, 21, and 33.
Accessed at http://darwin-online.org.uk/contents.html.

16. Darwin, *On the Origin of Species*, 74 and 169.

17. Here Darwin is explicitly referring to Henri Milne-Edwards's theory of physiological division of labor in organisms.

18. George W. Stocking, *Victorian Anthropology* (New York: Free Press, 1987); Michael
Adas, *Machines as the Measure of Men: Science, Technology, and Ideologies of Western
Dominance* (Ithaca, NY: Cornell University Press, 1989).

19. See Henry M. Cowles, "A Victorian Extinction: Alfred Newton and the Evolution
of Animal Protection," *British Journal for the History of Science* 46 (2013): 695–714.

20. E.g., "The Extinction of Animals," *Times* (London), May 19. 1884, 6; J. Robson,
"The Extinction of Primroses," *Times* (London), April 21, 1886, 4; P. P. Fraser, "The
Threatened Extinction of the Great Skua," *Times* (London), February 14, 1891, 13;
"The Extermination of the Ant-Bear," *Times* (London), September 24, 1892, 9; H. A.
Bryden, "The Extermination of the African Elephant," *Times* (London), November
12, 1895, 10.

21. E.g., "Extinct English Animals," *New York Times*, April 10, 1881, 4; "Influence of
Man on Animals," *New York Times*, October 30, 1881, 11; "Entire Races Extinct: Ani-
mals That Have Disappeared in Recent Times," *New York Times*, March 19, 1882, 6;
"Animals Exterminated by Man," *New York Times*, May 23, 1883, 3; "Cause for the
Extinction of the Horses of the Post-Tertiary," *New York Times*, April 8, 1883, 12;
"An Almost Extinct Tribe: The Remnant of the Alabama Indians in Texas," *New
York Times*, February 27, 1893, 3; "The Extinction of the Beaver," *New York Times*,
November 14, 1897, 11.

22. See Jim Downs, *Sick from Freedom: African-American Illness and Suffering during
the Civil War and Reconstruction* (New York: Oxford University Press, 2012).

Chapter 3

1.   D. H. Lawrence, *Lady Chatterley's Lover* (New York: Alfred A. Knopf, 1928), 1.

2.   J. Edward Chamberlin and Sander L. Gilman, *Degeneration: The Dark Side of Progress* (New York: Columbia University Press, 1985), ix.

3.   Darwin, *Origin of Species*, (London: John Murray, 1859), 13.

4.   For an accessible overview, see Sean B. Carroll, *Endless Forms Most Beautiful: The New Science of Evo Devo* (New York: W. W. Norton, 2006).

5.   Cesare Lombroso, *L'uomo delinquente* (Milan: Ulrico Hoepli, 1876).

6.   On the use of degeneration arguments in US immigration debates, see Peter Schrag, *Not Fit for Our Society: Nativism and Immigration* (Berkeley: University of California Press, 2010).

7.   E. Ray Lankester, *Degeneration: A Chapter in Darwinism* (London: Macmillan, 1880), 32.

8.   The literature on the eugenics movement is vast. For the canonical view, see Daniel J. Kevles, *In the Name of Eugenics: Genetics and the Uses of Human Heredity*, (Cambridge, MA: Harvard University Press, 1995).

9.   Modris Eksteins, "History and Degeneration: Of Birds and Cages," in Chamberlin and Gilman, *Degeneration*, 13. See also John Roderick Hinde, *Jacob Burckhardt and the Crisis of Modernity*, (Montreal: McGill-Queen's University Press, 2000), 200 ff.

10.   Georges Cuvier, quoted in Ignatius Donnelly, *Ragnarok: The Age of Fire and Gravel* (New York: D. Appleton and Company, 1883), title page.

11.   Martin Rudwick, *Earth's Deep History: How It Was Discovered and Why It Matters* (Chicago: University of Chicago Press, 2014).

12.   See, for example, Morton Paley, "*The Last Man*: Apocalypse without Millennium," in Audrey A. Fisch, Anne K. Mellor, and Esther H. Schor, eds., *The Other Mary Shelley: Beyond Frankenstein* (New York: New York University Press, 1993).

13.   *Literary Gazette and Journal of Belles Lettres* (1826), 103; *Monthly Review* (1826), 333–35; *Blackwood's Edinburgh Magazine* (1827), 54. These reviews are cited in Paley, "The Last Man" 108.

14.   Shiel, *The Purple Cloud*, 145.

15.   An older but still valuable study of this period in the history of biology is Peter J. Bowler, *The Eclipse of Darwinism: Anti-Darwinian Evolutionary Theories in the Decades around 1900* (Baltimore: Johns Hopkins University Press, 1983).

16.   See David Sepkoski, *Rereading the Fossil Record: The Growth of Paleobiology as an Evolutionary Discipline* (Chicago: University of Chicago Press, 2012), ch. 5.

17.   Throughout the rest of this book I will use the terms "taxon" (singular) and "taxa" (plural) to refer to units of taxonomic hierarchy (e.g., species, genera, families). This is a generic term, and it does not specify a particular "rank" in Linnaean hierarchy, but rather is a shorthand term adopted by scientists to stand for a "unit of biodiversity" at any hierarchical level.

18.   Paul D. Brinkman, *The Second Jurassic Dinosaur Rush: Museums and Paleontology in America at the Turn of the Twentieth Century* (Chicago: University of Chicago Press, 2010); David R. Wallace, *The Bonehunters' Revenge: Dinosaurs, Greed and the Great-*

est Scientific Feud of the Gilded Age (Boston: Houghton Mifflin, 1999); Mark Jaffe, *The Gilded Dinosaur: The Fossil War between E. D. Cope and O. C. Marsh and the Rise of American Science* (New York: Crown, 2000).

19. Olivier Rieppel, "Karl Beurlen (1901–1985), Nature Mysticism, and Aryan Paleontology," *Journal of the History of Biology* 45 (2012): 271.

20. Stephen Jay Gould, foreword to Otto H. Schindewolf, *Basic Questions in Paleontology* (Chicago: University of Chicago Press, 1993), ix.

21. Wolf-Ernst Reif, "Deutschsprachige Paläontologie im Spannungsfeld Zwischen Makroevolutionstheorie und Neo-Darwinismus (1920–1950)," in *Die Entstehung der Synthetischen Theorie: Beitruage zur Geschichte der Evolutionsbiologie in Deutschland 1930–1950*, ed. T. Junker and E.-M. Engels (Berlin: Verlag für Wissenschaft und Bildung, 1999); Reif, "The Search for a Macroevolutionary Theory in German Paleontology," *Journal of the History of Biology* 19 (1986); Reif, "Evolutionary Theory in German Paleontology," in Marjorie Glicksman Grene, ed., *Dimensions of Darwinism: Themes and Counterthemes in Twentieth-Century Evolutionary Theory* (Cambridge: Cambridge University Press, 1983).

Chapter 4

1. This story has been repeated, in some version or other, in most of the literature on the Manhattan Project. Its origin seems to be an interview Arthur Compton gave in 1959. See Pearl Buck, "The Bomb: The End of the World?" *American Weekly*, March 8, 1959. The story about bets being taken first appeared in Stephanie Groueff, *Manhattan Project: The Untold Story of the Making of the Atomic Bomb* (Boston: Little, Brown, 1967), 132. In 1991 Hans Bethe gave an interview clarifying the event and the actual risk associated with the test. See John Horgan, "Bethe, Teller, Trinity and the End of Earth," *Scientific American* (published online August 4, 2015), https://blogs.scientificamerican.com/cross-check/bethe-teller-trinity-and-the-end-of-earth/.

2. Oppenheimer made this statement in the 1965 television documentary *The Decision to Drop the Bomb*. The clip can be viewed at http://www.atomicarchive.com/Movies/Movie8.shtml.?.

3. See, for example, Paul Erickson et al., *How Reason Almost Lost Its Mind: The Strange Career of Cold War Rationality* (Chicago: University of Chicago Press, 2013).

4. On apocalyptic and postapocalyptic genres, see Frank Kermode, *The Sense of an Ending: Studies in the Theory of Fiction* (Oxford, UK: Oxford University Press, 2000 [1967]); and Teresa Heffernan, *Post-Apocalyptic Culture: Modernism, Postmodernism, and the Twentieth-Century Novel* (Toronto: University of Toronto Press, 2008).

5. Lyotard's book was originally commissioned as a report by the University of Québec, and published in French in 1979. It was published in translation in 1984. Jean-François Lyotard, *The Postmodern Condition: A Report on Knowledge* (Minneapolis: University of Minnesota Press, 1984).

6. The literature on Postmodernism is huge. For an overview, see Fredric Jameson,

Postmodernism; or, the Cultural Logic of Late Capitalism (Durham, NC: Duke University Press, 1992).

7. On Modernist literature, see Michael Levenson, ed., *The Cambridge Companion to Modernism* (Cambridge: Cambridge University Press, 1999).

8. On millennialism, see Frederic J. Baumgartner, *Longing for the End: A History of Millennialism in Western Civilization* (London: Palgrave Macmillan, 2001); and Catherine Wessinger, ed., *The Oxford Handbook of Millennialism* (Oxford, UK: Oxford University Press, 2011).

9. Weart, *Nuclear Fear*, 81–82.

10. Weart, *Nuclear Fear*, 78–79.

11. Karl Jaspers, *The Future of Mankind* (Chicago: University of Chicago Press, 1961), 5.

12. "Television," *The World Book Encyclopedia* (Chicago: World Book, 2003), 119.

13. Harlow Shapley to Macmillan Company, January 18, 1950; http://www.varchive .org/.

14. Michael Gordin's *The Pseudo-Science Wars* also comprehensively documents the furor around the book's publication, including peer and publication reviews.

15. The "millions" assertion comes from Stevin Shapin, "Catastrophism," *London Review of Books* 34 (November 8, 2012), 35.

16. See Gordin, *Pseudo-Science Wars*, 32ff, for a detailed discussion of the reception.

17. Immanuel Velikovsky, *Worlds in Collision* (New York: Macmillan, 1950), 383.

18. David Sepkoski, *Rereading the Fossil Record: The Growth of Paleobiology as an Evolutionary Discipline* (Chicago: University of Chicago Press, 2012), ch. 1.

19. Sepkoski, *Rereading the Fossil Record*, ch. 2.

20. Newell, "Periodicity," 384.

21. M. W. de Laubenfels, "Dinosaur Extinction: One More Hypothesis," *Journal of Paleontology* 30 (1956), 207–18.

22. Allan O. Kelly and Frank Dachille, *Target: Earth; the Role of Large Meteors in Earth Science* (Pensacola Engraving, 1953); René Gallant, *Bombarded Earth: An Essay on the Geological and Biological Effects of Huge Meteorite Impacts* (London: J. Baker, 1964).

23. Trevor Palmer, *Controversy: Catastrophism and Evolution: The Ongoing Debate* (New York: Kluwer Academic / Plenum Publishers, 1999), 105.

24. The obscure publication where this translation appeared in 1977 was *Catastrophist Geology*, a heavily Velikovskian magazine published for a few years during the late 1970s. It mostly featured essays by amateur geologists on the fringes of academic science, but did occasionally feature serious articles such as contributions by Schindewolf and the British paleontologist Derek Ager. This article was originally published as Otto H. Schindewolf, "Neokatastrophismus?" *Deutsche Geologische Gesellschaft Zeitschrift* 114, no. 2 (1963): 430–45.

25. James R. Beerbower, *Search for the Past: An Introduction to Paleontology* (Englewood Cliffs, NJ: Prentice-Hall, 1960).

26. On the history of ecology, see Gregg Mitman, *The State of Nature: Ecology, Community, and American Social Thought, 1900–1950* (Chicago: University of Chicago Press, 1992); Joel Bartholemew Hagen, *An Entangled Bank: The Origins of Eco-*

system Ecology (New Brunswick, NJ: Rutgers University Press, 1992); Sharon E. Kingsland, *The Evolution of American Ecology, 1890–2000* (Baltimore: Johns Hopkins University Press, 2005).

27.   Robert E. Kohler, *Lords of the Fly: Drosophila Genetics and the Experimental Life* (Chicago: University of Chicago Press, 1994).

28.   Sharon E. Kingsland, *Modeling Nature: Episodes in the History of Population Ecology* (Chicago: University of Chicago Press, 1985); Nancy G. Slack, *G. Evelyn Hutchinson and the Invention of Modern Ecology* (New Haven and London: Yale University Press, 2010).

29.   Sharon E. Kingsland, "The Refractory Model: The Logistic Curve and the History of Population Ecology," *Quarterly Review of Biology* 57, no. 1 (1982); Sepkoski, *Rereading the Fossil Record.*, ch. 4.

30.   Robert A. MacArthur, "Fluctuations of Animal Populations and a Measure of Community Stability," *Ecology* 36 (1955): 535.

31.   Thomas Robertson, *The Malthusian Moment: Global Population Growth and the Birth of American Environmentalism* (New Brunswick, NJ: Rutgers University Press, 2012), xii.

32.   Robertson, *The Malthusian Moment*, 144.

33.   Jacob Darwin Hamblin, *Arming Mother Nature: The Birth of Catastrophic Environmentalism* (Oxford, UK: Oxford University Press, 2013).

34.   LeRoy Stegman, "The Ecology of the Soil," transcription of a seminar at the New York State University College of Forestry, 1960; quoted in Carson, *Silent Spring*, 61.

Chapter 5

1.   Figures provided by "Top 100 Rated TV Shows Of All Time," http://tvbythenumbers .zap2it.com/reference/top-100-rated-tv-shows-of-all-time/. Accessed August 28, 2017.

2.   Carl Sagan, "The Nuclear Winter: The World after Nuclear War," *Parade*, October 30, 1983; Carl Sagan, "Nuclear War and Climatic Catastrophe: Some Policy Implications," *Foreign Affairs* 62 (1983), 257–92; Richard P. Turco et al., "Nuclear Winter: Global Consequences of Multiple Nuclear Explosions, *Science* 222 (1983): 1283–92; Paul R. Ehrlich et al., "Long-Term Biological Consequences of Nuclear War," *Science* 222 (1983): 1293–1300.

3.   See, e.g., Robert J. Lieber and Dan Horowitz, "Live, Die: Moot Point," *New York Times*, November 20, 1983.

4.   Luis W. Alvarez et al., "Extraterrestrial Cause for the Cretaceous-Tertiary Extinction," *Science* 208, no. 4448 (1980). For the purposes of brevity, I will refer to the impact extinction scenario developed by Luis and Walter Alvarez, along with Frank Asaro and Helen V. Michel, as the "Alvarez hypothesis." While it is often referred to as such in the literature, it is important to stress—as Walter Alvarez has himself insisted—that all collaborators should be credited with the discovery.

5.   The Cretaceous period is shortened to "K" because of the distinctive chalk deposits found in many of its geological formations ("Kreide" is the German word for

"chalk"). As mentioned in the introduction, I have opted to use the more familiar, colloquial designation "Tertiary" to describe the geological period known officially as the "Paleogene."

6. David M. Raup to Thomas J. M. Schopf, 28 January, 1979. Schopf papers, box 3, folder 30.

7. The history discussed in this section is a much-condensed account of developments discussed in my previous book *Rereading the Fossil Record*. Those interested in the full story should consult chapters 8 and 9, in particular.

8. See, e.g., James W. Valentine, "Patterns of Taxonomic and Ecological Structure of the Shelf Benthos During Phanerozoic Time," *Palaeontology* 12 (1969): 684–709.

9. See J. John Sepkoski, Jr., *A Compendium of Fossil Marine Families* ([Milwaukee]: Milwaukee Public Museum, 1982); "What I Did with My Research Career; or How Research on Biodiversity Yielded Data on Extinction," in *The Mass-Extinction Debates; How Science Works in a Crisis*, ed. William Glen (Stanford, CA: Stanford University Press, 1994).; Sepkoski, *Rereading*, chs. 7–10.

10. "A Kinetic Model of Phanerozoic Taxonomic Diversity: I. Analysis of Marine Orders," *Paleobiology* 4, no. 3 (1978): 223–51.

11. David M. Raup, *The Nemesis Affair: A Story of the Death of Dinosaurs and the Ways of Science* (New York: W. W. Norton, 1986), 52.

12. Walter Sullivan, "Two New Theories Offered on Mass Extinctions in Earth's Past," *New York Times*, June 10, 1980.

13. Leon T. Silver and Peter Schultz, preface to *Geological Implications of Impacts of Large Asteroids and Comets on the Earth*, Geological Society of America Special Paper 190 (1982), xi.

14. Interview with Walter Alvarez, June 8, 2007.

15. O. B. Toon et al., "Evolution of an Impact-Generated Dust Cloud and its Effects on the Atmosphere," in Silver and Schultz, eds., *Geological Implications of Impacts of Large Asteroids and Comets on the Earth* (Geological Society of America, 1982), 194–97.

16. J. John Sepkoski, Jr., "Mass Extinctions in the Phanerozoic Oceans: A Review," in Silver and Schultz, eds., *Geological Implications of Impacts*, 283–89; David M. Raup and J John Sepkoski, Jr., "Mass Extinctions in the Marine Fossil Record," *Science* 215, no. 4539 (1982): 1501–3.

17. Naomi Oreskes, *The Rejection of Continental Drift: Theory and Method in American Earth Science* (New York: Oxford University Press, 1999).

18. One of the most compelling, if also most difficult to establish, features of this particular period in the history of extinction is the influence of revolutionary politics and countercultural protest on scientists who participated in the debates. Stephen Jay Gould, for example, has made provocative statements implying that Marxism may have influenced his evolutionary views; but, despite his well-known leftist politics and membership in radical groups like Science for the People, he has staunchly denied that his science should be read through the lens of politics. Similarly, paleontologists like Jack Sepkoski and David Jablonski who were instrumental in revising the understanding of the role of mass extinctions in the history of

life had been active in protest movements and were lifelong fans of pop counterculture. Sepkoski, in particular, was an avid fan of punk rock music, and was even
described in his 1999 *New York Times* obituary—which otherwise focused on his
scientific accomplishments—as working in an office amid "the blaring sounds of
musical groups like the Sex Pistols." Carol Kaesuk Yoon, "J. John Sepkoski Jr., 50,
Dies; Changed Field of Paleontology," *New York Times*, May 6, 1999.

19. The first statement came during Reagan's address to the British Parliament on
June 8, 1982; Reagan used the phrase "evil empire" to describe the Soviet Union in
a number of contexts beginning in 1983.

20. Lewis Thomas, foreword to Paul R. Ehrlich et al., *The Cold and the Dark: The World
after Nuclear War* (New York: W.W. Norton, 1984), xxi–xxiii.

21. Donald Kennedy, introduction to Ehrlich et al., *The Cold and the Dark*, xxx.

22. Sharon Brownlee, "The Evidence: Cycles of Extinction," *Discover*, May 1984, 24;
"Did Comets Kill the Dinosaurs?," *Time*, May 6, 1985; Boyce Rensberger, "Extinction Governing Force in Theory of Evolution," *Washington Post*, November 24,
1984; Derek York, "Patterns of Mass Extinctions Not Just Chance, Theorists Say,"
Toronto Globe and Mail, July 29, 1985.

23. Raup, *Nemesis Affair*, 114.

24. Elizabeth S. Clemens, "The Impact Hypothesis and Popular Science," in William
Glen, ed., *The Mass Extinction Debates*, 114.

25. David Jablonski, "Causes and Consequences of Mass Extinctions; a Comparative
Approach," in *Dynamics of Extinction*, 183–230; Daniel Simberloff, "Are We on the
Verge of a Mass Extinction In the Tropics?" in *Dynamics of Extinction*, 165–80.

26. M. R. Rampino and R. B. Stothers, "Geological Rhythms and Cometary Impacts,"
Science 226 (1984): 1427–31.

27. See, e.g., "A Death-Star Theory Is Born," *Newsweek*, March 5, 1984.

28. "Miscasting the Dinosaurs' Horoscope," *New York Times*, April 2, 1985.

29. Niles Eldredge and Stephen Jay Gould, "Punctuated Equilibria: An Alternative to
Phyletic Gradualism," in *Models in Paleobiology* (San Francisco: Freeman, Cooper
& Co., 1972); Stephen Jay Gould and Niles Eldredge, "Punctuated Equilibria:
The Tempo and Mode of Evolution Reconsidered," *Paleobiology* 3, no. 2 (1977):
115–51.

30. Thomas S. Kuhn, *The Structure of Scientific Revolutions* (Chicago: University of Chicago Press, 1962).

31. Jean Baudrillard, "The Anorexic Ruins," in Dietmar Kamper and Christoph Wulf,
eds., *Looking Back on the End of the World* (New York: Semiotext(e), 1989), 29 and
33.

32. Michel de Montaigne, quoted in Schell, "The Second Death," 78.

Chapter 6

1. Walter G. Rosen to E. O. Wilson, June 23, 1992. Edward O. Wilson papers, box 141.

2. E. O. Wilson, *The Diversity of Life* (Cambridge, MA: Belknap Press, 1992); United
Nations, "Convention on Biological Diversity," 1992.

3. Elizabeth Kolbert's best-selling, Pulitzer Prize–winning 2014 book capitalized on, but did not create, the wide currency of the "sixth extinction" label. Elizabeth Kolbert, *The Sixth Extinction: An Unnatural History* (New York: Henry Holt, 2014).

4. The current concept of the "Anthropocene," which will be discussed further in the epilogue, argues that the human impact on the natural environment is so significant that it will be detectible in the earth's strata millions of years from now, and thus should be recognized as a new geological epoch. While neither the International Commission on Stratigraphy nor the International Union of Geophysical Sciences—the governing bodies that approve changes to stratigraphic designations—have approved this change, the term has entered broader usage in the environmental sciences, the social sciences, and the arts and humanities to describe the perception that humans have crossed an irreversible threshold due to anthropogenic climate change, pollution, and biodiversity loss.

5. E. O. Wilson, ed., *Biodiversity* (Washington: National Academy Press, 1988), v.

6. This statement was reported in, for example, "Scientists See Signs of Mass Extinction," *Washington Post*, September 29, 1986.

7. On Mayr's role in establishing the Modern Synthesis, see, e.g., Joseph A. Cain, "Ernst Mayr as Community Architect: Launching the Society for the Study of Evolution and the Journal Evolution," *Biology and Philosophy* 9 no. 3 (1994): 387–427; V. B. Smocovitis, "Unifying Biology: The Evolutionary Synthesis and Evolutionary Biology," *Journal of the History of Biology* 25, no. 1 (1992): 1–65; Ernst Mayr, *The Growth of Biological Thought: Diversity, Evolution, and Inheritance* (Cambridge, MA: Belknap Press, 1982).

8. Ernst Mayr, "The Diversity of Life," in Hadler et al., *Biology and the Future of Man*, ed. Handler Philip (Oxford, UK: Oxford University Press, 1970), 525.

9. There is a large scientific literature on this topic. For a short summary of the arguments, see P. Balvanera, G. C. Daily, P. R. Ehrlich, T. H. Ricketts, S. Bailey, S. Kark, C. Kremen, and H. Pereira, "Conserving Biodiversity and Ecosystem Services," *Science* 291 (2001): 2047.

10. On the history of the establishment of the influence of statistical analyses on public affairs, see Theodore M. Porter, *Trust in Numbers: The Pursuit of Objectivity in Science and Public Life* (Princeton, NJ: Princeton University Press, 1995).

11. Indeed, E. O. Wilson has written compellingly about what he calls "biophilia," which he regards as an essential ingredient for pursuing the study of nature. Edward O. Wilson, *Biophilia* (Cambridge, MA: Harvard University Press, 1984). In autobiographical accounts written by biologists from Charles Darwin to Richard Dawkins, similar motivations feature prominently. See David Sepkoski, "Two Lives in Biology," *Quarterly Review of Biology* 89 (2014): 151–56.

12. Information about Myers's career is taken from a variety of documents in the Edward O. Wilson papers, including letters exchanged with Wilson and an undated CV (most likely from the mid-1980s). Edward O. Wilson Papers, box 21.

13. Myers to Wilson, October 16, 1976; Wilson to Myers, November 12, 1976. Edward O. Wilson papers, United States Library of Congress Archives, box 21.

14. Wilson to Myers, July 19, 1983. Wilson papers, box 21.

15. Wilson to Myers, December 30, 1991. Wilson papers, box 59. E. O. Wilson, "Resolutions for the 80s," *Harvard Magazine* (January-February 1980), 21.

16. Wilson to Charlotte Mayerson, February 26, 1981. Wilson Papers, box 64.

17. Ehrenfeld, *Arrogance of Humanism*, 133.

18. Wilson to Soule, August 31, 1984. Wilson papers, box 156.

19. E. O. Wilson, "The Biological Diversity Crisis," *Bioscience* 35 (December 1985): 700-706.

20. E. O. Wilson, "The Biological Diversity Crisis: A Challenge to Science," *Issues in Science and Technology* 2 (1985): 20-29. This issue was published in the fall of 1985.

21. Wilson to Philip S. Cook, June 14, 1985. Wilson papers, box 248.

22. Wilson to Peter H. Raven, June 10, 1985. Wilson papers, box 139.

23. Wilson, "The Biological Diversity Crisis," 702-3.

24. These calculations are my own reconstruction of Erwin's figures, since Erwin himself does not spell them out. They appear to be accurate, however, since they produce roughly the number of species Erwin predicts.

25. Raven to Wilson, June 6, 1986. Wilson papers, box 139.

26. Wilson to Raven, June 10, 1986. Wilson papers, box 139.

27. Wilson, *Biodiversity*. The articles were included in *Science* 253 (August 16, 1991): 5021.

28. William A. Berggren and John A. Van Couvering, eds., *Catastrophes and Earth History: The New Uniformitarianism* (Princeton, NJ: Princeton University Press, 1984).

29. In his book *Time's Arrow, Time's Cycle*, Stephen Jay Gould reflected at length on the contrasting cultural understandings of time at scales of both natural and human history. Stephen Jay Gould, *Time's Arrow, Time's Cycle: Myth and Metaphor in the Discovery of Geological Time* (Cambridge, MA: Harvard University Press, 1987).

30. For a more detailed examination of the history of extinction research during this period, see Sepkoski, *Rereading the Fossil Record*, ch. 9.

31. Michael J. Bean to E. O. Wilson, December 11, 1981. E. O. Wilson Papers, Box 65.

32. See, e.g., Michael J. Benton, *When Life Nearly Died: The Greatest Mass Extinction of All Time* (London and New York: Thames and Hudson, 2003); Peter Douglas Ward, *Rivers in Time: The Search for Clues to Earth's Mass Extinctions* (New York: Columbia University Press, 2000); Douglas H. Erwin, *Extinction: How Life on Earth Nearly Ended 250 Million Years Ago* (Princeton, NJ, Princeton University Press, 2006); Norman MacLeod, *The Great Extinctions: What Causes Them and How They Shape Life* (London: Natural History Museum, 2013).

33. David M. Raup and J. John Sepkoski Jr., "Mass Extinctions in the Marine Fossil Record," *Science* 215, no. 4539 (1982).

34. Raup and Sepkoski, "Mass Extinctions in the Marine Fossil Record."; G. R. McGhee Jr., P. M. Sheehan, D. J. Bottjer, and M. L. Droser, "Ecological Ranking of Phanerozoic Biodiversity Crises: Ecological and Taxonomic Severities Are Decoupled," *Palaeogeography, Palaeoclimatology, Palaeoecology* 211 (2004): 289-97.

35. For a good (though somewhat partisan) general overview of these debates, see Anthony Hallam and Paul B. Wignall, *Mass Extinctions and Their Aftermath* (Oxford, UK: Oxford Science Publications, 1997).

36. Norman Myers, "The End of the Lines?" *Natural History* (February 1985), 2–12.

37. Norman Myers, "Mass Extinctions: What Can the Past Tell Us about the Present and Future?" *Palaeogeography, Palaeoclimatology, Palaeoecology* 82 (1990): 176.

38. Interview with David Jablonski, February 27, 2017. See, e.g., David Jablonski, "Extinctions: A Paleontological Perspective," *Science* 253, no. 10 (1991): 757; Sepkoski, "Biodiversity; Past, Present, and Future." Sepkoski thanked Myers for "incisive comments" on his article.

39. David M. Raup to E. O. Wilson, June 5, 1986. E. O. Wilson papers, box 139.

40. E. O. Wilson to David M. Raup, June 10, 1986. E. O. Wilson papers, box 139.

41. David M. Raup to E. O. Wilson, September 8, 1990. E. O. Wilson papers, box 139.

42. Norman Myers to E. O. Wilson, October 10, 1984; Norman Myers to E. O. Wilson, June 20, 1985. E. O. Wilson papers, box 121.

43. Norman Myers to E. O. Wilson, April 3, 1991. E. O. Wilson papers, box 121.

44. Norman Myers to E. O. Wilson, July 5, 1994. E. O. Wilson papers, box 121.

45. Myers to Wilson, July 5, 1994.

46. Norman Myers to David Jablonski, July 12, 1994. Courtesy of David Jablonski.

47. Norman Myers, "The Biodiversity Crisis and the Future of Evolution," *The Environmentalist* 16 (1995): 37–47. See David Jablonski, "The Tropics as a Source of Evolutionary Novelty through Geological Time," *Nature* (London) 364, no. 6433 (1993): 142–44.

48. Interview with David Jablonski, February 27, 2017.

49. For example, Farnham contends that during the second half of the twentieth century, "diversity emerged as a normative good, and this cultural development likely contributed to the popularity of diversity in environmental circles." Farnham, *Saving Nature's Legacy*, 7.

50. To be very clear, I am not endorsing these arguments. I am merely pointing out how their logic bears on current valuations of diversity.

Epilogue

1. Peter Wignall, *The Worst of Times: How Life on Earth Survived Eighty Million Years of Extinctions* (Princeton, NJ: Princeton University Press, 2015); Michael J. Benton, *When Life Nearly Died: The Greatest Mass Extinction of All Time* (London: Thames & Hudson, 2003).

2. Shu-Zhong Shen et al., "A Sudden End Permian Mass Extinction in South China," *GSA Bulletin* 131 (2018): 205–23.

3. In fact, just as this book was being completed, the Intergovernmental Science-Policy Platform on Biodiversity and Ecosystem Services (IPBES), an independent intergovernmental body closely associated with the United Nations, announced a major new report on anthropogenic threats to biological diversity. The report itself, "Global Assessment Report on Biodiversity and Ecosystem Services," is billed as "the most comprehensive ever completed," and was composed by well over a hundred scientists representing dozens of countries (https://www.ipbes.net /news/Media-Release-Global-Assessment). A central finding of the report is that

as many as a million species are currently threatened with extinction, which the report estimates to comprise an astounding 25 percent of all well-studied living groups. While the report presents nothing that is strikingly new to observers of biodiversity decline, it is quite noteworthy for both its comprehensiveness (the final document is more than 1,500 pages long and claims to synthesize the findings of more than 15,000 scientific sources) and the unanimity of the scientific opinion it represents: the IPBES claims more than 130 nations as members, and representatives from more than 50 of those countries contributed to the report.

4.   It should be stressed that this idea of a "neoliberal" Anthropocene is far from the only perspective on the impending challenges of climate change and social inequality. Nor are the voices in this conversation exclusively white, male, and Western. Indeed, a number of scholars in a variety of fields have explored the ways in which the current "culture" of the Anthropocene is implicated in histories of capitalism, colonialism, and racism. The geographer Kathryn Yusoff has, for example, highlighted the role of "extractive economies" of slavery and colonialism in producing the ecology of the Anthropocene in her 2019 book *A Billion Black Anthropocenes or None* (Minneapolis: University of Minnesota Press, 2019). Likewise, Jason Moore has reframed the debate around the concept of the "Capitalocene," which, in the introduction to a recent collection, he argues focuses on "questions of capitalism, power and class, anthropocentrism, dualist framings of 'nature' and 'society,' and the role of states and empire," which are "frequently bracketed by the dominant Anthropocene perspective." Jason Moore, ed., *Anthropocene or Capitalocene? Nature, History, and the Crisis of Capitalism* (Oakland, CA: PM Press, 2016), 5. This perspective is echoed in a collaborative project at the University of Wisconsin organized by an interdisciplinary group of scholars around the "Plantationocene"; the project's aim is "to come to terms with the plantation as a transformational moment in human and natural history on a global scale that is at the same time attentive to structures of power embedded in imperial and capitalist formations, the erasure of certain forms of life and relationships in such formations, and the enduring layers of history and legacies of plantation capitalism that persist, manifested in acts of racialized violence, growing land alienation, and accelerated species loss" (https://humanities.wisc.edu/research/plantationocene). Finally, in an effort to account not only for a diverse array of human perspectives on environmental change, but also for *nonhuman* ones, the critical theorist Donna Haraway has proposed the term "Chthulucene" to describe the "mixed assemblages" of human and nonhuman refugees of climate disaster. Donna Haraway, "Anthropocene, Capitalocene, Plantationocene, Chthulucene: Making Kin," *Environmental Humanities* 6 (2015): 159–65.

WORKS CITED

Adams, Brooks. 1896. *The Law of Civilization and Decay: An Essay on History*. New York: Macmillan.

Adas, Michael. 1989. *Machines as the Measure of Men: Science, Technology, and Ideologies of Western Dominance*. Ithaca, NY: Cornell University Press.

Abel, Othenio. 1921. "Gedanken über die Ursachen der Degeneration und deren phlogenetische Bedeutung," *Palaeontologia Hungarica* 1.

"A Death-Star Theory is Born." 1984. *Newsweek*, March 5.

Alvarez, Luis W., et al. 1980. "Extraterrestrial Cause for the Cretaceous-Tertiary Extinction." *Science* 208, no. 4448: 1095–1108.

Alvarez, Walter. 1997. *T. Rex and the Crater of Doom*. Princeton, NJ: Princeton University Press.

———. 2007. Interview with Walter Alvarez, June.

Amrine, Michael. 1950. "What the Atomic Age Has Done to Us." *New York Times*, August 6.

"An Almost Extinct Tribe: The Remnant of the Alabama Indians in Texas." 1893. *New York Times*, February 27.

"Animals Exterminated by Man." 1883. *New York Times*, May 23.

Balvanera, Patricia, et al. 2001. "Conserving Biodiversity and Ecosystem Services." *Science* 291, no. 5511: 2047.

Barrow, Mark V. 2009. *Nature's Ghosts: Confronting Extinction from the Age of Jefferson to the Age of Ecology*. Chicago: University of Chicago Press.

Baudrillard, Jean. 1989. "The Anorexic Ruins." In *Looking Back on the End of the World*, eds. Dietmar Kamper and Christoph Wulf. New York: Semiotext(e).

Baumgartner, Frederic J. 2001. *Longing for the End: A History of Millennialism in Western Civilization*. London: Palgrave Macmillan.

Beck, Ulrich. 1992 [1986]. *Risk Society: Towards a New Modernity*. London: Sage Publications.

Beerbower, James R. 1960. Search for the Past: An Introduction to Paleontology. Englewood Cliffs, NJ: Prentice-Hall.

Benton, Michael J. 2003. *When Life Nearly Died: The Greatest Mass Extinction of All Time*. London and New York: Thames and Hudson.

Berggren, William A. and John A Van Couvering, eds. 1984. *Catastrophes and Earth History: The New Uniformitarianism*. Princeton, NJ: Princeton University Press.

Beurlen, Karl. 1932. "Funktion und Form in der organischen Entwicklung." *Die Naturwissenschaften* 20, no. 5: 73–80.

———. 1937. *Die Stammesgeschichtlichen Grundlagen der Abstammungslehre*. Jena: Fischer.

Bowler, Peter J. 1976. "Malthus, Darwin, and the Concept of Struggle." *Journal of the History of Ideas* 37:631–50.

———. 1983. *The Eclipse of Darwinism: Anti-Darwinian Evolutionary Theories in the Decades around 1900*. Baltimore: Johns Hopkins University Press.

———. 1996. *Life's Splendid Drama: Evolutionary Biology and the Reconstruction of Life's Ancestry, 1860–1940*. University of Chicago Press.

Brannen, Peter. 2017. "When Life on Earth was Nearly Extinguished." *New York Times*, July 29.

Brantlinger, Patrick. 2003. *Dark Vanishings: Discourse on the Extinction of Primitive Races, 1800–1930*. Ithaca, NY: Cornell University Press.

Brinkman, Paul D. 2010a. "Charles Darwin's Beagle Voyage, Fossil Vertebrate Succession, and "The Gradual Birth & Death of Species." *Journal of the History of Biology* 43, no. 2: 363–99.

———. 2010b. *The Second Jurassic Dinosaur Rush: Museums and Paleontology in America at the Turn of the Twentieth Century*. Chicago: University of Chicago Press.

Broughton, Archdeacon. 1836. "Report from the Select Committee on Aborigines." British Parliamentary Papers, 1836.

Brownlee, Sharon. 1984. "The Evidence: Cycles of Extinction." *Discover*, May.

Bryden, H. A. 1895. "The Extermination of the African Elephant." *Times* (London), November 12.

Büchner, Fredrich Karl Christian Ludwig. 1875. *Man in the Past, Present, and Future*. London: Asher & Co.

Buck, Pearl. 1959. "The Bomb: The End of the World?" *American Weekly*, March 8.

Byron, Lord. 1816. "Darkness."

Cain, Joseph A. 1994. "Ernst Mayr as Community Architect: Launching the Society for the Study of Evolution and the Journal Evolution." *Biology and Philosophy* 9, no. 3: 387–427.

Campbell, Thomas. 1823. "The Last Man."

Carroll, Sean B. 2006. *Endless Forms Most Beautiful: The New Science of Evo Devo*. New York: W. W. Norton.

Carson, Rachel. 1962. *Silent Spring*. Cambridge MA: Riverside Press.

Carter, Jimmy. 1980. "Preface to Council on Environmental Quality." In *The Eleventh*

Annual Report of the Council on Environmental Quality. Washington: Executive Office of the President, Council on Environmental Quality.

Cassirer, Ernst. 1946. *The Myth of the State.* New Haven: Yale University Press.

"Cause for the Extinction of the Horses of the Post-Tertiary." 1883. *New York Times,* April 8.

Chamberlin, J. Edward, and Sander L. Gilman. 1985. *Degeneration: The Dark Side of Progress.* New York: Columbia University Press.

Clemens, Elizabeth S. 1994. "The Impact Hypothesis and Popular Science." in *The Mass Extinction Debates: How Science Works in a Crisis,* edited by William Glen. Stanford, CA: Stanford University Press.

Congressional Research Service. 1982. *A Legislative History of the Endangered Species Act of 1973, as amended in 1976, 1977, 1978, 1979, and 1980.* Washington: Government Publishing Office.

Corson, Eugene Rollin. 1893. *The Future of the Colored Race in the U.S.* Ithaca, NY: Wilder Quarter Century Book.

Cottrell, Leonard S., and Sylvia Eberhart. 1969 [1948]. *American Opinion on World Affairs in the Atomic Age.* New York: Greenwood Press.

Council of Environmental Quality. 1980. *Eleventh Annual Report of the Council on Environmental Quality.* Washington: Executive Office of the President, Council on Environmental Quality.

Cousins, Norman. 1945. "Modern Man is Obsolete." *Saturday Review,* August 18.

Cowles, Henry M. 2013. "A Victorian Extinction: Alfred Newton and the Evolution of Animal Protection." *British Journal for the History of Science* 46:694–714.

Crutzen, Paul J. 2002. "Geology of Mankind." *Nature* 3, no. 415: 23.

Cuvier, Georges. 1997. "Revolutions of the Surface of the Globe." In Georges Cuvier, *Fossil Bones and Geological Catastrophes,* translated and edited by Martin J. S. Rudwick. Chicago: University of Chicago Press.

Damuth, John. 1992. "Extinction." In *Keywords in Evolutionary Biology,* edited by Evelyn Fox Keller and Elisabeth A. Lloyd. Cambridge, MA: Harvard University Press.

Darwin, Charles. N.d. "February 1835." DAR42.97–99. Transcribed and edited by John van Wyhe. Darwin Online, http://darwin-online.org.uk/.

———. 1839. *Journal of Researches.* London: John Murray.

———. 1845. *Journal of Researches.* London: John Murray.

———. 1859. *Origin of Species.* London: John Murray.

———. 1871. *The Descent of Man, and Selection in Relation to Sex.* London: John Murray.

———. 1872. *On the Origin of Species,* 6th edition. London: John Murray.

———. 1887. "Autobiography." In *The Life and Letters of Charles Darwin,* edited by Francis Darwin London: John Murray.

———. 1909. "Essay of 1842" and "Essay of 1844." In *The Foundation of the Origin of Species,* edited by Francis Darwin. Cambridge: Cambridge University Press.

———. 1909. *The Voyage of the Beagle.* New York: P. F. Collier & Son.

———. 1987. *Charles Darwin's Notebooks, 1836–1844: Geology, Transmutation of Species, Metaphysical Enquiries*. London and Ithaca, NY: British Museum (Natural History) and Cornell University Press.

Davis, Doug. 2001. "'A Hundred Million Hydrogen Bombs': Total War in the Fossil Record." *Configurations* 9, no. 3: 461–508.

Davis, Marc, Piet Hut, and Richard A. Muller. 1984. "Extinction of Species by Periodic Comet Showers." *Nature* 308:715–17.

"The Decision to Drop the Bomb." 1965. National Broadcasting Company.

"Did Comets Kill the Dinosaurs?" 1985. *Time*, May 6.

Dobzhansky, Theodosius. 1937. *Genetics and the Origin of Species*. New York: Columbia University Press.

———. 1962. *Mankind Evolving: The Evolution of the Human Species*. New Haven and London: Yale University Press.

Dominici, Stefano. 2010. "Brocchi's Subapennine Fossil Conchology." *Evolution: Education, and Outreach* 3, no. 4: 585–94.

Donnelly, Ignatius. 1883. *Ragnarok: The Age of Fire and Gravel*. New York: D. Appleton and Company.

Downs, Jim. 2012. *Sick from Freedom: African-American Illness and Suffering during the Civil War and Reconstruction*. New York: Oxford University Press.

Ehrenfeld, David W. 1972. *Conserving Life on Earth*. 1972. Oxford, UK: Oxford University Press.

———. 1978. *The Arrogance of Humanism*. Oxford, UK: Oxford University Press.

Ehrlich, Paul. 1968. *The Population Bomb*. New York: Ballantine Books.

———. 1986. "Extinction: What is Happening Now and What Needs to Be Done." in *Dynamics of Extinction*, edited by David K. Elliot. New York: John Wiley and Sons.

Ehrlich, Paul, and L. C. Birch. 1967. "The 'Balance of Nature' and 'Population Control.'" *The American Naturalist* 101, no. 918 (March-April): 97–107.

Ehrlich, Paul R., and Anne Ehrlich. 1970. *Population, Resources, Environment: Issues in Human Ecology*. San Francisco: W. H. Freeman.

———. 1981. *The Causes and Consequences of the Disappearance of Species*. New York: Random House.

Ehrlich, Paul, et al. 1983. "Long-Term Biological Consequences of Nuclear War." *Science* 222:1293–1300.

———. 1984. *The Cold and the Dark: The World After Nuclear War*. New York: W. W. Norton.

Eksteins, Modris. 1985. "History and Degeneration: Of Birds and Cages." In *Degeneration: The Dark Side of Progress*, edited by J. Edward Chamberlain and Sander L. Gilman. New York: Columbia University Press.

Eldredge, Niles. 1991. *The Miner's Canary: Unraveling the Mysteries of Extinction*, 1st ed. New York: Prentice Hall.

———. 1977. "Punctuated Equilibria: The Tempo and Mode of Evolution Reconsidered." *Paleobiology* 3, no. 2: 115–51.

———. 2009. "Experimenting with Evolution: Darwin, the Beagle, and Evolution." *Evolution: Education, and Outreach* 2, no. 1: 35–54.

Eldredge, Niles, and Stephen Jay Gould. 1972. "Punctuated Equilibria: An Alternative to Phyletic Gradualism." In *Models in Paleobiology*. San Francisco: Freeman, Cooper & Co.

Eliot, T. S. 1925. "The Hollow Men." *Poems 1909–1925*. London: Faber & Gwyer.

Elton, Charles. 1930. *Animal Ecology and Evolution*. Oxford, UK: Oxford University Press.

———. 1958. *The Ecology of Invasions by Animals and Plants*. London: Methuen Publishing.

———. 1982. "Entire Races Extinct: Animals That Have Disappeared in Recent Times." *New York Times*, March 19.

Erikson, Kai. 1982. "A Horror beyond Comprehension." *New York Times*, April 11.

Erickson, Paul, et al. 2013. *How Reason Almost Lost Its Mind: The Strange Career of Cold War Rationality*. Chicago: University of Chicago Press.

Erwin, Douglas H. 2006. *Extinction: How Life on Earth Nearly Ended 250 Million Years Ago*. Princeton, NJ: Princeton University Press.

Erwin, Terry L. 1982. "Tropical Forests: Their Richness in Coleoptera and Other Arthropod Species." *Coleopterists Bulletin* 36, no. 1: 74–75.

"The Extermination of the Ant-Bear." 1892. *Times* (London), September 24. "Extinct English Animals." 1881. *New York Times*, April 10.

"The Extinction of Animals." 1884. *Times* (London), May 19.

"The Extinction of the Beaver." 1987. *New York Times,* November 14, 11.

Farnham, Timothy J. 2007. *Saving Nature's Legacy: Origins of the Idea of Biological Diversity*. New Haven: Yale University Press.

———. 2016. "A Confluence of Values: Historical Roots of Concern for Biological Diversity." In *The Routledge Handbook of Philosophy of Biodiversity*, edited by Justin Garson, Anya Plutynski, and Sahotra Sarkar. London: Routledge.

Fischer, Alfred G., and Michael A. Arthur. 1977. "Secular Variations in the Pelagic Realm." In *Society of Economic Paleontologists and Mineralogists Special Publication 25* (November): 19–50.

Flessa, Karl W., and John Imbrie. 1973. "Evolutionary Pulsations: Evidence from Phanerozoic Diversity Patterns." In *Implications of Continent Drift to the Earth Sciences, Vol. 1*, edited by D. H. Tarling and S. K. Runcorn. London and New York: Academic Press.

———. 1984. "Extinctions Are In." *Paleobiology* 12, no. 3: 329–34.

Fletcher, Amy Lynn. 2014. *Mendel's Ark: Biotechnology and the Future of Extinction*. New York: Springer.

Foucault, Michel. 1994 [1966]. *The Order of Things: An Archaeology of the Human Sciences*. New York: Vintage.

Fraser, P. P. 1891. "The Threatened Extinction of the Great Skua." *Times* (London), February 14.

Fukuyama, Francis. 1989. "The End of History?" *The National Interest* 16:3–18.

Gallant, René. 1964. *Bombarded Earth: An Essay on the Geological and Biological Effects of Huge Meteorite Impacts*. London: J. Baker.

Galton, Francis. 1869. *Hereditary Genius: An Inquiry into Its Laws and Consequences*. London: Nabu Press.

Goodman, Ellen. 1984. "Musings of a Dinosaur Groupie." *Washington Post*, January 3.

Gordin, Michael D. 2012. *The Pseudoscience Wars: Immanuel Velikovsky and the Birth of the Modern Fringe*. Chicago: University of Chicago Press.

Gould, Stephen Jay. 1984a. "Sex, Drugs, Disasters, and the Extinction of Dinosaurs." *Discover*, May.

———. 1984b. "The Cosmic Dance of Siva." *Natural History* 93, no. 8 (August).

———. 1984c. "Toward the Vindication of Punctuational Change." In *Catastrophes and Earth History: The New Uniformitarianism*, edited by W. A. Berggren and A. Van Couvering John. Princeton, NJ: Princeton University Press.

———. 1987. *Time's Arrow, Time's Cycle: Myth and Metaphor in the Discovery of Geological Time*. Cambridge, MA: Harvard University Press.

———. 1993. Foreword to *Basic Questions in Paleontology*. By Otto H. Schindewolf. Chicago: University of Chicago Press.

Gould, Stephen Jay, and Niles Eldredge. 1977. "Punctuated Equilibria: The Tempo and Mode of Evolution Reconsidered." *Paleobiology* 3, no. 2: 115–51.

Greg, William R. 1868. "On the Failure of 'Natural Selection' in the Case of Man." *Fraser's Magazine*, September.

Groueff, Stephanie. 1967. *Manhattan Project: The Untold Story of the Making of the Atomic Bomb*. Boston: Little, Brown.

Groves, Craig, et al. 2002. "Planning for Biodiversity Conservation: Putting Conservation Science into Practice," *BioScience* 52, no. 6: 499–512.

Haeckel, Ernst. 1866. *Generelle Morphologie der Organismen*, vol. 2. Berlin: G. Reimer.

Hagen, Joel. 1992. *An Entangled Bank: The Origins of Ecosystem Ecology*. New Brunswick, NJ: Rutgers University Press.

Hallam, Anthony, and Paul B. Wignall. 1997. *Mass Extinctions and Their Aftermath*. Oxford, UK: Oxford Science Publications.

Haller, John S. 1971. *Outcasts from Evolution: Scientific Attitudes of Racial Inferiority, 1859–1900*. Urbana: University of Illinois Press.

Hamblin, Jacob Darwin. 2013. *Arming Mother Nature: The Birth of Catastrophic Environmentalism*. Oxford, UK: Oxford University Press.

Haraway, Donna. 2015. "Anthropocene, Capitalocene, Plantationocene, Chthulucene: Making Kin," *Environmental Humanities* 6:159–65.

Hartmann, William K., and Donald R. Davis. 1975. "Satellite-Sized Planetesimals and Lunar Origin." ICARUS 24:504–15.

"Hearings Open on Endangered Species Law." 1981. *New York Times*, December 9.

Heffernan, Teresa. 2008. *Post-Apocalyptic Culture: Modernism, Postmodernism, and the Twentieth-Century Novel*. Toronto: University of Toronto Press.

Heise, Ursula. 2016. *Imagining Extinction: The Cultural Meanings of Endangered Species*. Chicago: University of Chicago Press.

Hill, David L., Eugene Rabinowitch, and John A. Simpson. 1945. "The Atomic Scientists Speak Up." *Life*, October 29.

Hinde, John Roderick. 2000. *Jacob Burckhardt and the Crisis of Modernity*. Montreal: McGill-Queen's University Press.

Hobsbawm, E. J. 1989. *The Age of Empire, 1875–1914*. New York: Vintage.

Hobson, John A. 1902. "The Scientific Basis of Imperialism." *Political Science Quarterly* 17, no. 3: 460–61.

Horgan, John. 2015. "Bethe, Teller, Trinity and the End of Earth." *Scientific American*, August 4.

Hutchinson, G. Evelyn. 1959. "Homage to Santa Rosalia; or, Why Are There So Many Kinds of Animals?" *American Naturalist* 93, no. 870 (May-June): 145–59.

"Influence of Man on Animals." 1881. *New York Times*, October 30.

Jablonski, David.

———. 1986a "Background and Mass Extinctions: The Alternation of Macroevolutionary Regimes." *Science* 231, no. 4734: 129–33.

———. 1986b. "Causes and Consequences of Mass Extinctions: A Comparative Approach." In *Dynamics of Extinction.*, edited by K. Elliott David. New York: John Wiley and Sons.

———. 1986c. "Mass Extinction; New Answers, New Questions," in *The Last Extinction*, edited by Les Kaufman and Kenneth Mallory. Cambridge: MIT Press.

———. 1991. "Extinctions: A Paleontological Perspective." *Science* 253, no. 5021 (August): 754–57.

———. 1993. "The Tropics as a Source of Evolutionary Novelty through Geological Time." *Nature* (London) 364, no. 6433.

———. 1994. "Extinctions in the Fossil Record." In *Estimating Extinction Rates: Sir Joseph Banks Anniversary Meeting; a Discussion*, edited by J. H. Lawton and R. M. May, Philosophical Transactions of the Royal Society of London, Series B: Biological Sciences. London: Royal Society of London.

———. 2017. Interview with David Jablonski. 27 February.

Jaffe, Mark. 2000. *The Gilded Dinosaur: The Fossil War between E. D. Cope and O. C. Marsh and the Rise of American Science*. New York: Crown Publishing Group.

Jaher, Frederic Cople. 1964. *Doubters and Dissenters: Cataclysmic Thought in America, 1885–1918*. London: Free Press of Glencoe.

Jameson, Fredric. 1992. *Postmodernism; or, The Cultural Logic of Late Capitalism*. Durham, NC: Duke University Press, 1992.

Jaspers, Karl. 1961. *The Future of Mankind*. Chicago: University of Chicago Press.

Jefferson, Thomas. 1801. *Notes on the State of Virginia*, Eighth American ed. Newark: Pennington and Gould.

Kahn, Herman. 1960. *On Thermonuclear War*. Princeton, NJ: Princeton University Press.

Kazin, Alfred. 1950. "On the Brink." *New Yorker*, April 29.

Kelly, Allan O., and Frank Dachille. 1953. *Target: Earth: The Role of Large Meteors in Earth Science*. Pensacola, FL: Pensacola Engraving.

Kennedy, David. 1984. Introduction to *The Cold and the Dark: The World after Nuclear War*, by Paul R. Ehrlich. New York: W. W. Norton.

Kermode, Frank. 2000. *The Sense of an Ending: Studies in the Theory of Fiction*. Oxford, UK: Oxford University Press.

Kevles, Daniel J. 1995. *In the Name of Eugenics: Genetics and the Uses of Human Heredity*. Cambridge, MA: Harvard University Press.

Kingsland, Sharon E. 1982. "The Refractory Model: The Logistic Curve and the History of Population Ecology." *Quarterly Review of Biology* 57, no. 1: 29–52.

———. 1985. *Modeling Nature: Episodes in the History of Population Ecology, Science and Its Conceptual Foundations*. Chicago: University of Chicago Press.

———. 2005. *The Evolution of American Ecology, 1890–2000*. Baltimore: Johns Hopkins University Press.

Kohler, Robert E. 1994. *Lords of the Fly: Drosophila Genetics and the Experimental Life*. Chicago: University of Chicago Press.

Kolbert, Elizabeth. 2014. *The Sixth Extinction: An Unnatural History*. New York: Henry Holt.

Kuhn, Thomas S. 1962. *The Structure of Scientific Revolutions*. Chicago: University of Chicago Press.

Kump, Lee. 2018. "Climate Change and Marine Mass Extinction." *Science* 362, no. 6419: 1113–14.

Lamarck, Jean Baptiste de, and Pierre Antoine de Monet. 1984. *Zoological Philosophy: An Exposition with Regard to the Natural History of Animals*. Chicago: University of Chicago Press.

Lankester, E. Ray. 1880. *Degeneration: A Chapter in Darwinism*. London: Macmillan.

Larrabee, Eric. 1950. "The Day the Sun Stood Still." *Harper's Magazine*, January.

Laubenfels, M. W. 1956. "Dinosaur Extinction: One More Hypothesis." *Journal of Paleontology* 30:207–12.

Lawrence, D. H. 1928. *Lady Chatterley's Lover*. New York: Alfred A. Knopf.

Leakey, Richard, and Roger Lewin. 1995. *The Sixth Extinction: Patterns of Life and the Future of Humankind*. New York: Doubleday.

Le Conte, Joseph. 1892. *The Race Problem in the South*. New York: D. Appleton.

Levenson, Michael, ed. 1999. *The Cambridge Companion to Modernism*. Cambridge and New York: Cambridge University Press.

Lieber, Robert J. and Dan Horowitz. 1983. "Live, Die: Moot Point." *New York Times*, November 20.

Linnaeus, Carolus. 1762. "The Oeconomy of Nature." In *Miscellaneous Tracts Relating to Husbandry and Physick*. London: J. Dodsley.

Livingstone, David N. 2008. *Adam's Ancestors: Race, Religion, and the Politics of Human Origins*. Baltimore: Johns Hopkins University Press.

Lombroso, Cesar. 1876. *L'uomo delinquente*. Milan: Ulrico Hoepli.

London, Jack. 1915. *The Scarlet Plague*. New York: Macmillan.

Lovejoy, Arthur O. 2009. *The Great Chain of Being: A Study of the History of an Idea.* New Brunswick, NJ: Transaction Publishers.

Ludwig, Büchner, Friedrich Karl Christian. 1872. *Man in the Past, Present, and Future.* London: Asher & Co.

Lull, Richard Swann. 1917. *Organic Evolution.* New York: Macmillan.

Lyell, Charles. 1826. "Review of Transactions of the Geological Society." *Quarterly Review* 24:507–40.

———. 1830–33. *Principles of Geology; Being an Attempt to Explain the Former Changes of the Earth's Surface, by Reference to Causes Now in Operation.* London: J. Murray.

Lyotard, Jean-François. 1984. *The Postmodern Condition: A Report on Knowledge.* Minneapolis: University of Minnesota Press.

MacArthur, Robert A. 1955. "Fluctuations of Animal Populations and a Measure of Community Stability." *Ecology* 36, no. 3 (July): 533–36.

MacLaurin, James, and Kim Sterelny. 2008. *What Is Biodiversity?* Chicago: University of Chicago Press.

MacLeod, Norman. 2013. *The Great Extinctions: What Causes Them and How They Shape Life.* London: Natural History Museum.

Maffi, Luisa, ed. *On Biocultural Diversity: Linking Language, Knowledge, and the Environment.* 2001. Washington: Smithsonian Institution Press.

Malpas, Simon. 2005. *The Postmodern.* New York: Routledge.

Malthus, Thomas. 1826. *Essay on the Principle of Population*, 6th edition. London John Murray.

Mayr, Ernst. 1970. "The Diversity of Life," in *Biology and the Future of Man*, edited by Handler Philip. Oxford, UK: Oxford University Press.

———. 1982. *The Growth of Biological Thought: Diversity, Evolution, and Inheritance.* Cambridge, MA: Belknap Press.

———. 1988. "How Many Species Are There on Earth?" *Science* 241, no. 4872 (September): 1441–49.

Maza, Sarah. 2003. *The Myth of the French Bourgeoisie: An Essay on the Social Imaginary, 1750–1850.* Cambridge, MA: Harvard University Press.

McCausland, James. 2006. "Scientists' Warnings Unheeded." *Courier-Mail*, December 4.

McKibben, Bill. 1989. *The End of Nature.* New York: Penguin.

———. 2014. "Postscript: Jonathan Schell." *New Yorker*, April 7.

"Miscasting the Dinosaurs' Horoscope." 1985. *New York Times*, April 2.

Mitman, Gregg. 1992. *The State of Nature: Ecology, Community, and American Social Thought, 1900–1950, Science and Its Conceptual Foundations.* Chicago: University of Chicago Press.

———. 2018. "Hubris or Humility? Genealogies of the Anthropocene." In *Future Remains: A Cabinet of Curiosities for the Anthropocene*, edited by Gregg Mitman, Marco Armiero, and Robert S. Emmett, 59–68. Chicago: University of Chicago Press.

Moore, Jason. 2016. *Anthropocene or Capitalocene? Nature, History, and the Crisis of Capitalism*. Oakland, CA: PM Press.

Murry, George, to George Arthur. 1830. Reprinted in "Van Diemen's Land. Copies of All Correspondence between Lieutenant-Governor Arthur and His Majesty's Secretary of State for the Colonies, on the Subject of the Military Operations Largely Carried on Against the Aboriginal Inhabitants of Van Diemen's Land," British Parliamentary Papers, 1831.

Myers, Norman. 1979. *The Sinking Ark: A New Look at the Problem of Disappearing Species*. New York: Pergamon Press.

———. 1985. "The End of the Lines." *Natural History* 94, no. 2: 2, 6, 12.

———. 1990. "Mass Extinctions: What Can the Past Tell Us about the Present and Future?" *Palaeogeography, Palaeoclimatology, Palaeoecology* 82, no. 1: 175–85.

———. 1996. "The Biodiversity Crisis and the Future of Evolution." *Environmentalist* 16:37–47.

Nettle, Daniel, and Suzanne Romaine. 2000. *Vanishing Voices: The Extinction of the World's Languages*. Oxford, UK: Oxford University Press.

Newell, Norman D. 1952. "Periodicity in Invertebrate Evolution." *Journal of Paleontology* 26, no. 3: 371–85.

———. 1956. "Catastrophism and the Fossil Record." *Evolution* 10, no. 1 (March): 97–101.

———. 1963. "Crises in the History of Life." *Scientific American* 208, no. 2: 76–93.

———. 1967. "Revolutions in the History of Life," in *Uniformity and Simplicity*. Boulder, CO: Geological Society of America.

Newton, Alfred. 1885. "Mr. Grieve on the Garefowl." *Nature* 32:545–46.

Nordau, Max Simon. 1895. *Degeneration: Translated from the Second Edition of the German Work*. London: W. Heinemann.

Ohendorf, Pat. 1983. "Catastrophes That Changed the World." *MacLean's*, December 26.

Oreskes, Naomi. 1999. *The Rejection of Continental Drift: Theory and Method in American Earth Science*. New York: Oxford University Press.

Osborn, Henry Fairfield. 1910. *Age of Mammals*. London: Macmillan.

Overbye, Denis. 1984. "The Theories: Cosmic Winter." *Discover*, May.

Packard, Alpheus. 1886. "Geological Extinction and Some of Its Apparent Causes." *American Naturalist* 20:29–40.

Paley, Morton. 1993. "The Last Man: Apocalypse without Millennium," in *The Other Mary Shelley: Beyond Frankenstein*, edited by Audrey A. Fisch, Anne K. Mellor, and Esther H. Schor. New York: New York University Press.

Palmer, Trevor. 1999. *Controversy: Catastrophism and Evolution, the Ongoing Debate*. New York: Kluwer Academic / Plenum Publishers.

Parkinson, James. 1804. *An Examination of the Mineralized Remains of the Vegetables and Animals of the Antediluvian World*, vol. 1. London: C. Whittingham.

Paul, Diane. 2003. "Darwin, Social Darwinism, and Eugenics." In *The Cambridge*

Companion to Darwin, edited by Jonathan Hodge and Gregory Radick. Cambridge: Cambridge University Press.

Pearce, Trevor. 2010. "'A Great Complication of Circumstances': Darwin and the Economy of Nature." *Journal of the History of Biology* 43, no. 3 (August): 493–528.

Pearson, Karl. 1905. *National Life from the Standpoint of Science*. Second Edition. London: Adam and Charles Black.

Penn, Justin, et al. 2018. "Temperature-Dependent Hypoxia Explains Biogeography and Severity of End-Permian Marine Mass Extinction." *Science* 362, no. 6419: 1130.

Porter, Theodore M. 1995. *Trust in Numbers: The Pursuit of Objectivity in Science and Public Life*. Princeton, NJ: Princeton University Press.

Prichard, James Cowles. 1840. "On the Extinction of Human Races." *Edinburgh New Philosophical Journal* 28: 166–70.

Purdy, Jedediah. 2015. "Anthropocene Fever." *Aeon*, March 31.

———. 2018. *After Nature: A Politics for the Anthropocene*. Cambridge, MA: Harvard University Press.

Rampino, Michael R., and Richard B. Stothers. 1984. "Geological Rhythms and Cometary Impacts." *Science* 226:1427–31.

Raup, David M. 1978. "Approaches to the Extinction Problem." *Journal of Paleontology* 52, no. 3: 517–23.

———. 1976. "Species Diversity in the Phanerozoic: A Tabulation." *Paleobiology* 2, no. 4: 279–88.

———. 1982. "Large Body Impacts and Terrestrial Evolution Meeting, October 19–22, 1981." *Paleobiology* 8, no. 1: 1–3.

———. 1986. *The Nemesis Affair: A Story of the Death of Dinosaurs and the Ways of Science*. New York: W. W. Norton.

———. 1991. *Extinction: Bad Genes or Bad Luck?* New York: W.W. Norton.

Raup, David M., and J John Sepkoski, Jr. 1982. "Mass Extinctions in the Marine Fossil Record." *Science* 215, no. 4539: 1501–3.

———. 1984. Periodicity of Extinctions in the Geologic Past." *Proceedings of the National Academy of Sciences of the United States of America* 81, no. 3: 801–5.

Reif, Wolf-Ernst. 1983. "Evolutionary Theory in German Paleontology," in *Dimensions of Darwinism: Themes and Counterthemes in Twentieth Century Evolutionary Theory*, edited by Marjorie Glicksman Grene. Cambridge: Cambridge University Press.

———. 1986. "The Search for a Macroevolutionary Theory in German Paleontology." *Journal of the History of Biology* 19, no. 1: 79–130.

———. 1999. "Deutschsprachige Paläontologie im Spannungsfeld zwischen Makroevolutionstheorie und Neo-Darwinismus (1920–1950)." In *Die Entstehung der Synthetischen Theorie: Beitruage zur Geschichte der Evolutionsbiologie in Deutschland 1930–1950*, edited by T. Junker and E.-M. Engels. Berlin: Verlag für Wissenschaft und Bildung.

Rensberger, Boyce. 1984. "Extinction Governing Force in Theory of Evolution." *Washington Post*, November 24.

Richards, Robert J. 2009. "Darwin's Theory of Natural Selection and Its Moral Purpose." In *The Cambridge Companion to the Origin of Species*, edited by Michael Ruse and Robert J. Richards. Cambridge: Cambridge University Press.

Rieppel, Olivier. 2012. "Karl Beurlen (1901–1985), Nature Mysticism, and Aryan Paleontology." *Journal of the History of Biology* 45, no. 2: 253–99.

Robertson, Thomas. 1886. *The Malthusian Moment: Global Population Growth and the Birth of American Environmentalism*. New Brunswick, NJ: Rutgers University Press.

Robson, J. 1886. "The Extinction of Primroses." *Times* (London), April 21.

Rudwick, Martin J. S. 2004. *Bursting the Limits of Time: The Reconstruction of Geohistory in the Age of Revolution*. Chicago: University of Chicago Press.

———. 2014. *Earth's Deep History: How It Was Discovered and Why It Matters*. Chicago: University of Chicago Press.

Sagan, Carl. 1983a. "ABC News Debate." Aired November 20.

———. 1983b. "Nuclear War and Climatic Catastrophe: Some Policy Implications." *Foreign Affairs* 62 (1983): 257–92.

———. 1983c. "The Nuclear Winter: The World after Nuclear War." *Parade*, October 30.

Sakar, Sahotra. 2002. *Biodiversity and Environmental Philosophy*. Cambridge: Cambridge University Press.

Shabecoff, Philip. 1981. "Hearings Open on Endangered Species Law," *New York Times*, December 9.

Schell, Jonathan. 1982. "The Second Death." *New Yorker*, February 8.

Schindewolf, Otto H. 1963. "Neokatastrophismus?" *Deutsche Geologische Gesellschaft Zeitschrift* 114, no. 2: 430–45. Translated and reprinted by V. Axel Firsoff, *Catastrophist Geology* 2, no. 2 (1977).

———. 1964. "Erdgeschichte und Weltgeschichte," Akademie der Wissenschaften und der Literatur. Abhandlungen der mathematisch-naturwissenschaftlichen Klasse. Mainze: Akademie d. Wiss. u.d. Literatur: Wiesbaden: Steiner in Komm.

Schmidt, Oscar. 1875. *The Doctrine of Descent and Darwinism*. London: H. S. King & Co.

Schrag, Peter. 2010. *Not Fit for Our Society: Nativism and Immigration*. Berkeley: University of California Press.

Schuchert, Charles. 1924. *A Text-Book of Geology*, 3rd ed., New York: Wiley and Sons.

Schwägerl, Christian. 2014. *The Anthropocene: The Human Era and How It Shapes Our Planet*. Santa Fe: Synergistic Press.

"Scientists See Signs of Mass Extinction." 1986. *Washington Post*, September 29.

Scranton, Roy. 2015. *Learning to Die in the Anthropocene: Reflections on the End of Civilization*. San Francisco: City Lights.

Selous, Frederick. 1896. *Sunshine and Storm in Rhodesia*, 2nd ed. London: Rowland Ward and Company.

Sepkoski, David. 2012. *Rereading the Fossil Record: The Growth of Paleobiology as an Evolutionary Discipline*. Chicago: University of Chicago Press.

———. 2014. "Two Lives in Biology." *Quarterly Review of Biology* 89: 151–56.

Sepkoski, J. John, Jr. 1978. "A Kinetic Model of Phanerozoic Taxonomic Diversity I. Analysis of Marine Orders." *Paleobiology* 4, no. 3: 223–51.

———. 1981. "A Factor Analytic Description of the Phanerozoic Marine Fossil Record." *Paleobiology* 7:36–53.

———. 1982a. *A Compendium of Fossil Marine Families*. Milwaukee: Milwaukee Public Museum.

———. 1982b. "Mass Extinctions in the Phanerozoic Oceans: A Review." In *Geological Implications of Impacts of Large Asteroids and Comets on the Earth*, edited by Leon T. Silver and Peter H. Schultz. Geological Society of America.

———. 1985. "Some Implications of Mass Extinction for the Evolution of Complex Life." In *The Search for Extraterrestrial Life: Recent Developments*, edited by M. D. Papagiannis. Dordrecht, Netherlands: D. Reidel.

———. 1994. "What I Did with My Research Career; or, How Research on Biodiversity Yielded Data on Extinction." In *The Mass-Extinction Debates; How Science Works in a Crisis.*, edited by William Glen. Stanford, CA: Stanford University Press.

———. 1997. "Biodiversity; Past, Present, and Future." *Journal of Paleontology* 71, no. 4: 533–39.

Shapin, Stevin. 2012. "Catastrophism." *London Review of Books*, November 8.

Sheehan, P. M., D. J. Bottjer, and M. L. Droser. 2004. "Ecological Ranking of Phanerozoic Biodiversity Crises: Ecological and Taxonomic Severities Are Decoupled." *Palaeogeography, Palaeoclimatology, Palaeoecology* 211:289–97.

Shen, Shu-Zhong. 1993. *Basic Questions in Paleontology*. Chicago: University of Chicago Press.

———. 2018. "A Sudden End Permian Mass Extinction in South China." *GSA Bulletin* 131:205–23.

Shiel, M. P. 1901. *The Purple Cloud*. London: Chatto & Windus.

Shils, Edward. 1956. *The Torment of Secrecy: The Background and Consequences of American Security Policies*. London: William Heinemann.

Silver, Leon T. and Peter Schultz. 1982. *Geological Implications of Impacts of Large Asteroids and Comets on the Earth*. Geological Society of America.

Simberloff, Daniel. 1987. "Are We on the Verge of a Mass Extinction in the Tropics?" *PALAIOS* 2, no. 2: 165–71.

Simpson, George Gaylord. 1944. *Tempo and Mode in Evolution*. New York: Columbia University Press.

———. 1949. *The Meaning of Evolution*. New Haven: Yale University Press.

Skutnabb-Kangas, Tove, Luisa Maffi, and David Harmon. 2003. *Sharing a World of Difference: The Earth's Linguistic, Cultural, and Biological Diversity*. Paris: UNESCO, Terralingua, and the World Wide Fund for Nature.

Slack, Nancy G. 2010. *Evelyn Hutchinson and the Invention of Modern Ecology*. New Haven and London: Yale University Press.

Smith, Charles Hamilton. 1851. *The Natural History of the Human Species: Its Typical Forms, Primeval Distribution, Filiations, and Migrations*. Boston: Gould and Lincoln.

Smith, Eric A. 2001. *On the Coevolution of Cultural, Linguistic, and Biological Diversity*. Washington and London: Smithsonian Institution Press.

Smocovitis, V. B. 1992. "Unifying Biology: The Evolutionary Synthesis and Evolutionary Biology." *Journal of the History of Biology* 25, no. 1: 1–65.

Soddy, Frederick. 1903. "Some Recent Advances in Radioactivity." *Contemporary Review* 83: 708–20.

Soulé, Michael. 1986. "What Is Conservation Biology?" *Bioscience* 35, no. 11 (December): 727–34.

———. 1987. "History of the Society for Conservation Biology: How and Why We Got Here." *Conservation Biology* 1, no. 1: 4–5.

Soulé, Michael, and Brian Wilcox. 1976. "Species Diversity in the Phanerozoic: An Interpretation." *Paleobiology* 2, no. 4 (Fall): 279–88.

———. 1980. "Conservation Biology: Its Scope and Challenge." In *Conservation Biology: An Evolutionary-Ecological Perspective*, edited by Soulé and Wilcox. Sunderland, MA: Sinauer, 1980.

Spencer, Herbert. 1851. *Social Statistics; or, The Conditions Essential to Happiness Specified, and the First of them Developed*. London: John Chapman.

———. 1852. "A Theory of Population, Deduced from the General Law of Animal Fertility." *Westminster Review* 57: 468–501.

———. 1873. *The Study of Sociology*. London: Henry S. King.

Spengler, Oswald. 1926. *The Decline of the West, Vol. 1*. New York: Alfred A. Knopf.

Staum, Martin S. 2003. *Labeling People: French Scholars on Society, Race, and Empire, 1815–1848*. Montreal: McGill-Queen's University Press.

Steffen, Will, et al. 2007. "The Anthropocene: Are Humans Now Overwhelming the Great Forces of Nature?" *Ambio* 36, no. 8 (December): 614–21.

Stocking, George W. 1987. *Victorian Anthropology*. New York: Free Press.

Sullivan, Walter. 1980. "Two New Theories Offered on Mass Extinctions in Earth's Past." *New York Times*, June 10.

———. 1982. "Mass Extinctions Increasingly Blamed on Catastrophes from the Sky." *New York Times*, January 19.

Takacs, David. 1996. *The Idea of Biodiversity: Philosophies of Paradise*. Baltimore: Johns Hopkins University Press.

"Television." 2003. *World Book Encyclopedia*. Chicago: World Book.

Thomas, Lewis. 1984. Foreword to *The Cold and the Dark: The World after Nuclear War*, by Paul R. Ehrlich. New York: W. W. Norton.

Thoreau, Henry David. 1854. *Walden*. Borton: Ticknor and Fields.

Toon, O. B., et al. 1982. "Evolution of an Impact-Generated Dust Cloud and Its Effects on the Atmosphere." In *Geological Implications of Impacts of Large Asteroids and Comets on the Earth*, edited by Leon T. Silver and Peter Schultz. Geological Society of America.

Turco, Richard P., et al. 1983. "Nuclear Winter: Global Consequences of Multiple Nuclear Explosions." *Science* 222:1283–92.

UNESCO. 2002. "UNESCO Universal Declaration on Cultural Diversity." Paris.

United Nations. 1992. "Convention on Biological Diversity."

Valentine, James W. 1969. "Patterns of Taxonomic and Ecological Structure of the Shelf Benthos during Phanerozoic Time." *Palaeontology* 12, part 4: 608–709.

Velikovsky, Immanuel. 1950. *Worlds in Collision*. New York: Macmillan and Company.

———. 1955. *Earth in Upheaval*. New York: Doubleday.

Vernadsky, Vladamir Ivanovich. 1944. "Problems of Biogeochemistry II: On the Fundamental Matter-Energy Difference between the Living and Inert Bodies of the Biosphere." *Transactions of the Connecticut Academy of Arts and Sciences* 35:483–512.

Wallace, Alfred Russell. 1864. "The Origin of Human Races and the Antiquity of Man Deduced from the Theory of Natural Selection." *Journal of the Anthropological Society of London* 2:162–64.

Wallace, David R. 1999. *The Bonehunters' Revenge: Dinosaurs, Greed and the Greatest Scientific Feud of the Gilded Age*. Boston: Houghton Mifflin.

———. 2019. *The Uninhabitable Earth: Life after Warming*. New York: Tim Duggan Books.

Ward, Barbara, and René Dubos. 1972. *Only One Earth: The Care and Maintenance of a Small Planet*. New York: W. W. Norton.

Ward, Lester F. 1903. *Pure Sociology: A Treatise on the Origins and Spontaneous Development of Society*. New York: Macmillan.

Ward, Peter Douglas. 2000. *Rivers in Time: The Search for Clues to Earth's Mass Extinctions*. New York: Columbia University Press.

Weart, Spencer. 1988. *Nuclear Fear: A History of Images*. Cambridge, MA: Harvard University Press.

Wells, H. G. 1895. *The Time Machine*. New York: Henry Holt.

———. 1920. *Outline of History*. New York: Macmillan.

Wessinger, Catherine, ed. 2011. *The Oxford Handbook of Millennialism*. Oxford, UK: Oxford University Press.

Wignall, Peter. 2015. *The Worst of Times: How Life on Earth Survived Eighty Million Years of Extinctions*. Princeton, NJ: Princeton University Press.

Wilford, John Noble. 1983. "Study Indicates Extinctions Strike in Regular Intervals." *New York Times*, December 11.

Wilson, E. O. 1980. "Resolutions for the 80s." *Harvard Magazine*, January-February.

————. 1984. *Biophilia*. Cambridge, MA: Harvard University Press.

————. 1985. "The Biological Diversity Crisis: A Challenge to *Science*." *Issues in Science and Technology* 2:20–29.

————. 1988. *Biodiversity*. Washington: National Academy Press.

————. 1988. "The Diversity of Life." In *Earth '88: Changing Geographic Perspectives*. National Geographic Society.

————. 1991. "Biodiversity." *Science* 253 (August 16): 5012.

————. 1992. *The Diversity of Life*. Cambridge, MA: Belknap Press.

Wollock, Jeffrey. 2001. "Linguistic Diversity and Biodiversity: Some Implications for the Language Sciences." In *On Biocultural Diversity: Linking Language, Knowledge, and the Environment*, edited by Luisa Maffi. Washington: Smithsonian University Press.

Woodward, Arthur Smith. 1910. "Presidential Address to Section C." In *Report of the British Association for the Advancement of Science* 79:462–71.

Worster, Donald. 1994. *Nature's Economy: A History of Ecological Ideas*, 2nd ed. Cambridge and New York: Cambridge University Press.

Wyndham, John. 1955. *The Chrysalids*. London: Penguin Books.

Yoon, Carol Kaesuk. 1999. "J. John Sepkoski Jr., 50, Dies, Changed Fields of Paleontology." *New York Times*, May 6.

York, Derek. 1985. "Patterns of Mass Extinctions Not Just Chance, Theorists Say." *Globe and Mail* (Toronto), July 29.

Yusoff, Kathryn. 2019. *A Billion Black Anthropocenes or None*. Minneapolis: University of Minnesota Press.

INDEX

Page numbers in italic indicate figures.

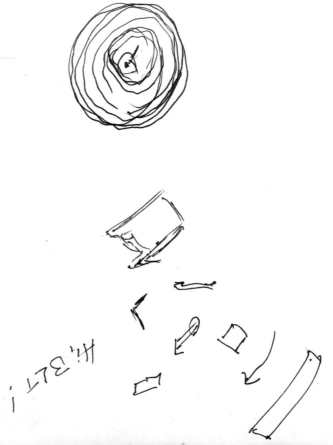

Hi, 347!